LEARNING with LEAN

Unleashing the Potential for Sustainable Competitive Advantage

LEARNING with LEAN

Unleashing the Potential for Sustainable Competitive Advantage

James Zurn • Perry Mulligan

CRC Press
Taylor & Francis Group
Boca Raton London New York

CRC Press is an imprint of the
Taylor & Francis Group, an **informa** business

A PRODUCTIVITY PRESS BOOK

CRC Press
Taylor & Francis Group
6000 Broken Sound Parkway NW, Suite 300
Boca Raton, FL 33487-2742

Printed on acid-free paper
Version Date: 20130422

International Standard Book Number-13: 978-1-4665-7294-2 (Paperback)

Library of Congress Cataloging-in-Publication Data

Zurn, James.
Learning with lean : unleashing the potential for sustainable competitive advantage / James Zurn, Perry Mulligan.
pages cm
Includes bibliographical references and index.
ISBN 978-1-4665-7294-2 (pbk.)
1. Lean manufacturing. 2. Quality control. I. Mulligan, Perry. II. Title.

HD58.9.Z87 2013
658.4'01--dc23 2013010931

Visit the Taylor & Francis Web site at
http://www.taylorandfrancis.com

and the CRC Press Web site at
http://www.crcpress.com

This book is for our wives, Sheila and Sue, who gave us endless support, encouragement, and tireless feedback on our thoughts during its construction.

You were our sounding boards, thought coaches, and cheerleaders. Throughout the whole process, you helped keep us grounded in reality, and this work would never have come to life without you.

With all our love, and more thanks than you can ever imagine.

Contents

Preface

Why would you want to read this book? Well for one thing, there are not many like it. It deals with the gap between a good performing organization and a great performing organization, one that is change-hungry and demonstrates a sustainable competitive advantage. Today, the gap may be concealed in the organization because your customers are generally happy with what they receive, stakeholders are happy with organizational performance, and your operational execution is pretty good. No matter how good the organization is performing, unless you have a sustainable competitive advantage you are at risk.

If that is the opportunity we are trying to address, what are the key elements on which the book is based that define our solution? As it relates to our leadership perspective and subsequently the contents of this book, two fundamental truths helped shape our thought process:

1. The only sustainable competitive advantage an organization has is the rate at which it learns, a derivative of the quote from Peter Senge (*The Fifth Discipline*, 1990), "The rate at which organizations learn may become the only sustainable source of competitive advantage."
2. Lean principles, defined as "a practice that considers the use of resources to achieve any goal other than the creation of value for the customer to be wasteful, and thus a target for elimination," apply to any process by providing opportunity to significantly transform and improve it.

What drove us into research and exploration of successful organizations' use of Lean as a learning vehicle to elevate operational performance? Moreover, why do we feel that the combination of the two will produce a sustainable competitive advantage? There are many well-written books on implementing the tools, techniques, and methods of Lean organizations. In most cases the focus of these books is organizations that either are just getting started or are early in their Lean journey. We applaud those efforts, but do they unlock the organization's potential to get to the next level of performance? No. Our focus is to provide a vehicle for organizations that are already successful, yet do not demonstrate a sustainable competitive advantage.

As is typical with most of our goals in life, "knowing" what to do is greatly different than being able to actually do it. We embraced the notion of sustainable learning, but did not fully understand how you could possibly achieve that within a typical workplace. The problem appeared to be a philosophical truth that had no practical application in the business world.

It was only through a more thorough understanding of what constituted the Lean journey that the ability to apply this philosophy came to life. Lean, in the drive to continuously eliminate waste, provides optimum solutions for your customers, and it provides the vehicle for applied continuous learning. By coupling these two concepts, we believe that it creates a strategy that lets you focus the organization's development, starting with the individual, on the path of continuous improvement using Lean as the common language to achieve its sustainable competitive advantage.

During the course of this book, it will lead you to understand what sustainable competitive advantage means to an organization, and how Lean best demonstrates itself as a vehicle for learning. By using a series of five simple phases, we will lead you from understanding your current state to developing an architecture and implementation plan to transform the organization's performance, deploy training and implement new knowledge and skills around Lean, and then physically find and improve opportunities where waste exists so that the organization can learn and reinvent the way it learns.

These five phases fit neatly into a closed-system model that has similarities to the Plan–Do–Check–Act quality model. The model is simple, easy to communicate, and easy to implement—Assess, Plan, Prepare, Do, and Learn.

Our work starts with a premise based on the assumptions about your organization today, which is already successful in its market against its competitors and has been for some time. Our research and experience find that while the successes of an organization's strength play well in the short term, they could potentially be one of the greatest limiting factors (lethargy and loss of agility) or even a significant threat to its long-term success (blind to disruptive technologies and market threats).

There is a premise built into the logic of this book; it centers on how organizations live, thrive, and fail based on the energy they use and the wastes they create to achieve their mission and deliver their outputs. It leads us to create thought-provoking topics aimed at shaking a successful organization free from its status quo performance, unlocking it from organizational cement that binds it tightly in one spot, and openly requires active dialogue about learning from both successes and failures. Good management techniques always recognize the need for balance, and there is nothing different in this book. There are no absolutes when it comes to establishing and fostering a learning culture with Lean, only shades of gray that come with attending levels of risk and rewards.

Our interest lies in helping you be able to recognize where those balances and dichotomies exist, and techniques you can use to understand and manage these tough challenges and ignite the fires within the organization to be change-hungry using Lean. The concept of deploying Learning with Lean throughout the organization is presented as the vehicle that can elevate an organization to a level of sustainable performance that will lead to a competitive advantage in its market, against its competitors, and establish it as a benchmark for execution.

Based on the five-phase model, we will spend time working on assessing and inventorying the 'as-is" of today's condition. Good information at this phase proves its worth and value as you move into developing a well-architected and crafted implementation plan. We spend only a small portion of time on Lean tools because our assumption is your Lean journey has started and you possess knowledge of how the tools and methods work. There will be time devoted to understanding the Voice of the Customer as a focusing engine, appropriate measurement and goal setting, organizational change dynamics, and learning how to learn with Lean.

We believe that organizations act and behave exactly like living organisms; that is, organizations mimic nature's biology. They use resources, exert energy to transform inputs into outputs, create heat and wastes, and must use feedback mechanisms to manage themselves for proper health. This book uses this premise in how it introduces methods and feedback systems, monitors performance, and manages output.

What we talk about throughout this book are good, commonsense actions. While the methods and techniques we use are simple, do not mistake the strategy as easy to deploy. The book provides the step-by-step methodology with emphasis on what should be used, when it should be used, and how it can be used to benefit moving an organization toward its best levels of performance.

This book is a recipe for transformation, how to start a journey, and the steps you should take as a leader to guide the organization toward becoming a change-hungry Lean Learning Organization positioned with a sustainable competitive advantage. It is aimed at giving senior leaders tools, methods, and structures that they can use to create a direction, chart the course for transformation, execute the journey of transformation, and lead their organization into being a sustainable Learning Organization that uses Lean as its learning engine.

Our intent is to use a conversational style and framework for the book. There will be many places where we have identified themes or topics and will talk about them in a matter-of-fact approach. By doing so, it allows us greater latitude in how to convey the messages and what they mean to you or your organization.

The purpose of this book is to outline the recipe for implementing the strategy, not as a reference manual for Lean principles and practices. It should be used as a leader's guide, an outline of how to assess, plan, prepare, and execute the strategy defined as Learning with Lean. It couples the two powerful concepts of sustainable competitive advantage and Lean as a learning tool together to accelerate your organization's rate of change. Learning is almost impossible without action and practice. You are investing time (your most valuable resource) in reading these messages. At the end of each chapter, you will be challenged to put into practice what you are reading. There are suggested actions where you should choose several to do that total at least 50 points. The first ones should be accomplished in the next few weeks, with those at the end of later chapters completed aggressively as you move from phase to phase.

If you do these earnestly, by the time you complete this book you will be solidly on your way toward transforming yourself and your organization into a Lean Learning Organization.

Our journey to create this book has been personally entertaining, thought provoking, and challenging for us. We sincerely hope that in reading this book you can unleash your own personal potential as a leader and enable your organization to excel.

Acknowledgments

We would like to recognize the following people who helped develop our perspectives, attitudes, and experiences along the way. Without their insight, mentoring, and challenging throughout the years we would not have been able to learn and practice our craft. Many of these friends and partners also kept a crisp eye on content and construction, and provided constructive feedback throughout the entire process. They helped us figure out how to convey the book's key messages and learnings.

In any work, it is hard to place people in an order of influence, impact, or contributions because each one brought something unique to its creation. Therefore, in our egalitarian style we have placed them in alphabetical order.

- David Fausel provided contributions throughout the production process and helped make it more readable, impactful, and added greater value to our readers.
- Chuck Gallenberg provided professional challenges during formative years that brought critical thinking into daily management work.
- Rudy Hacker has been a strong influence by providing a sense of consistency, purpose and urgency during the creation of improvement roadmaps in the organizations we touched together, while sharing common beliefs in values.
- Jay Margolis and his team wielded editorial wisdom and guidance during the creation process to greatly improve the final writing.
- Marc Onetto introduced me to the magical journey that is Lean and helped mold my concepts and understanding of its structure and nuances.
- Patricia Schill taught me to believe in myself.
- Michael Sinocchi found, promoted, and mentored initiation of the book and helped keep us on track during its creation.
- Karlheinz Totz coached me to always set the bar high, and to have faith that the team could deliver.

- John Vinyard has been a critical support during developmental years, and has always been a resilient sounding board during many years together on the Board of Examiners for NIST's Malcolm Baldrige National Quality Award program.
- James Zurn, Sr. gave tireless editorial reviews that helped improve the manuscript's correctness and readability.

Chapter 1

What Is the Need for Change?

> There is at least one point in the history of any organization when you have to change dramatically to rise to the next level of performance. Miss that moment and you start to decline.
>
> **Andrew S. Grove, 1936–present**
> *Engineer, Author, Business Leader*

Your organization is already achieving a level of success, and customers are generally pleased with the deliverables and operational performance. The value customers are receiving is reasonable compared to the current market, and your employees are generally satisfied and turnover is competitive with others.

- Why think about changing?
- What is the need, you are already successful?
- Why disrupt today's status quo—customers are pleased, employees are satisfied?
- Why consider this approach over others, all of which say almost the same thing?

All are good questions that raise challenges for the Learning with Lean (LWL) approach. The major difference is in the starting baseline—it is given that your organization is already performing reasonably well. Performance is good, customers are generally satisfied, your employee base is stable and trained on their work tasks, and financially the organization is in good condition.

Based on existing performance levels, questions about the future naturally form.

- How do you protect what exists today (market, customers, and technologies), while assuring it sustains and grows for tomorrow?
- How do you deliver better margins while achieving the mission?
- Is "customer delight" critical in the future versus having satisfied customers today?
- Are you sure that core competencies and skills will stay with the organization if new opportunities show up in the market?
- Are your technology, service, and intellectual property growing at a rate that assures future competitiveness?
- How do you have to deliver more effectively and efficiently without spending more to get it?

Throughout this book, we will use questions like these to explore and demonstrate why using Lean as a vehicle for Organizational Learning (OL) is vital to building core strength and creating a competitive position.

We have a premise to offer with a personal challenge. Examine the premise, determine the degree of your organization's fit to the premise, and candidly explore your desire and commitment to accelerate your organization toward greater productivity through becoming an organization that learns with them.

The premise—it is not about Lean (the methods and tools), but is about learning to become a Learning Organization. It uses simple Lean tools and practices as the core learning vehicle and change agent.

The challenge—are you ready to make a personal commitment to examine and identify opportunities in the organization, then use Lean as a vehicle to create a cultural change that ultimately achieves a sustainable competitive advantage?

1.1 What Is Sustainable Competitive Advantage?

> The rate at which organizations learn may become the only sustainable source of competitive advantage.
>
> **Peter Senge**

Achieving the ideal state reflected by a "Sustainable Competitive Advantage" is the Holy Grail for most organizations. Possessing that secret recipe that ensures long-term success has been the magic elixir that many have tried to find.

The purpose of this book is to remove the mystery from this process. It will demonstrate the recipe for accomplishing this goal by doing two things:

- Breaking down what is meant by a Sustainable Competitive Advantage.
- Constructing a recipe, a step-by-step formula that allows you to position your organization on the path of delivering this advantage, regardless of what product or service you provide.

For this purpose, a Sustainable Competitive Advantage is defined as:

> the ability to deliver a solution that the customer values in a way that is not available through other sources.

That differentiation can be in any one or combination of the following attributes: price, quality, features, serviceability, or ease of use. It is very important to remember that the attribute must be determined by the customer—what they perceive as value. The significance of the customer's vindication of the advantage will become clearer with discussions on threats that the organization may face. Being internally focused in the quest to differentiate and achieve a competitive advantage has led many organizations down a path that leverages their technology to deliver industry-leading performance that the customer does not need or see value in.

To augment the basic definition requires exploration of boundary conditions that will help understand what learning as a sustainable advantage looks like in the real world. First, let us explore the notion of pure research as the opportunity to create a Learning Organization. "Trickle-down" technology was a key benefit of the space program in the 1960s when many new technologies, materials, and systems were developed. However, does pure research offer an opportunity for any organization to achieve the necessary rate of learning? There are specific industries and service organizations where this form of academic endeavor will net the rate of learning but the belief is that these represent too small a population to consider this sufficient focus for most organizations.

1.1.1 Applied Rate of Learning

If theoretical research is not sufficient, will applying this research to the specific goods or services rendered be enough to drive the necessary Applied Rate of Learning (AROL)? The act of applying learning is a fundamental element of the formula to achieve the desired AROL.

For this book, the concept of AROL is extracted from concepts presented in *Combining Service and Learning, A Resource Book for Community and Public Service* and is defined as:

> a set of actions, or activities, that engage an organization and its employees directly with a phenomena (condition, opportunity or challenge) being examined and involve them in creating solutions that connect the phenomena with application of theoretical concepts.

In Figure 1.1, the significant shift in how changes and improvements occur is through alignment (clustering of small improvements) using common tools and methods, focusing on what is important to the organization, and ultimately raising the bar on the speed of learning and sharing of knowledge.

The act of applying learning in most cases engages more of the population of the organization. This is a case of where size does matter. The more people on the learning curve, the easier it is for the entire organization to hit

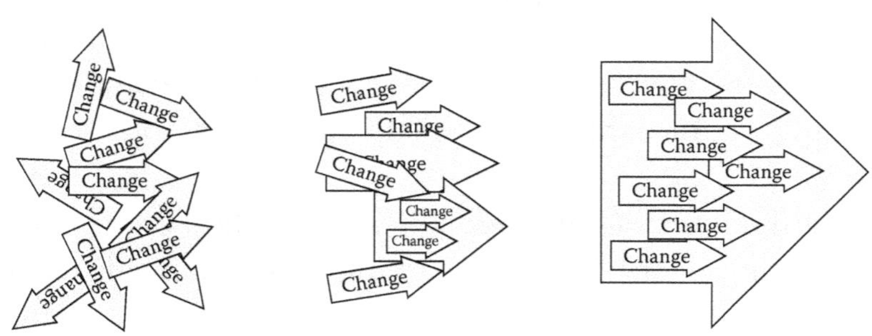

Organization getting smarter, but competency not applied toward same goal, no synergy evident, no learning in place

Changes happening, organization generally aligned, some synergy emerging, pockets of performance excellence and learning present

Systemic changes happen, synergy of improvements evident across organization, improvements matter for mission success and its customers satisfaction, individual and organizational learning in place

Figure 1.1 Applied Rate of Learning transformation.

the necessary AROL. This activity also exposes the organization to a necessary practice to drive increased AROL. The act of translating macro or theoretical concepts into specific practical applications is a practice that is very important in the process of increasing the AROL. The "macro–micro–macro" cycle is referred to as porpoising, which truly helps the broader audience understand the concepts and the direction of the organization.

Every person has a different, preferred method of learning. By spanning the scope of concept granularity, it will undoubtedly help many people in the organization understand what you are trying to accomplish in a way that makes it easier for them to accept and internalize the ideas, and then contribute to the overall deployment. The ability to improve the rate of adoption of new ideas is critical to achieving an improved AROL and porpoising contributes to this effort. So, is applied learning enough to achieve the desired state? Unfortunately, no; there is another critical factor that is missing—structure or content.

That is where the practices of Lean and its tools come into the equation. The basic principles of Lean—the elimination of waste and focus on meeting customers' expectations with the least resources consumed—are ideas that transcend any function, group, or individual within the organization. This fact, coupled with the fact that the application of Lean does not deliver an end solution but is rather a journey that is in essence never ending, makes Lean the perfect vehicle to be used to teach an organization how to apply learning. It will help you with a common language, focus, and set of criteria for success. Because of its universal applicability, it is equally usable in the boardroom as it is on the production floor. With the Lean learning vehicle in place, the focus on applying the learning across the entire organization combines with the passion for learning to achieve a sustainable competitive advantage. At this point, you have defined the critical factors necessary to deploy LWL as a strategic initiative to achieve this goal.

1.2 Threats

> The nature of the global organization environment guarantees that no matter how hard we work to create a stable and healthy organization, our organization will continue to experience dramatic changes far beyond our control.
>
> **Margaret J. Wheatley**

Now that the ideal end-game is defined, scrutiny should be applied against potential threats to the organization. More specifically, you need to answer the question of whether this initiative is neutralized by these threats or offers a solution to protect against them.

1.2.1 Leadership

When conducting a SWOT (Strength, Weakness, Opportunity, and Threat) (Panagiotou, 2003) analysis for the organization, an often overlooked element that needs to be considered is "management preference." Checking the alignment of management preference to any strategic direction is critical. If an organization wants to be fast and nimble, and the management preferences for decision making are autocratic and dictatorial, a misalignment is evident. If you look at the strategy of creating a Lean Learning Organization, there are several checks that should be applied to management preference to determine if this is a threat.

The two basic questions that need to be answered are:

1. Does management believe that the organization must change to remain successful in the future?
2. Does management understand and support the premise of this strategy?

If the answer to either of these questions is no, then the proposed strategy is a non-starter. As you look to evaluate this strategy against the other threats, consider this item neutral since you cannot proceed without the appropriate executive sponsorship, conviction, and support.

1.2.2 Organizational

Organizational structure, by definition, is a suboptimized solution to address conflicting requirements. The natural tension in the various elements of the organization tends to improve performance on one dimension, or discipline, while subjugating other functions. Within this diverse set of solutions, how does the suggested strategy fair? For this strategy to succeed, there are two organizational attributes that need to be present:

- executive sponsorship, and
- a deployment map at the single contributor level.

Since you have already considered the need for executive sponsorship under the leadership section, let us consider the deployment plan. The deployment plan can start from a single person, department, function, organization unit, or the entire entity. Since it is not limited to scale, functional responsibilities, or location in the value stream, there are no predetermined organizational factors required. On this dimension of threats, consider the organization as having no threat to the strategy.

1.2.3 Technology

When considering technology threats you should assess if the strategy is restricted to technology maturity, changes in process, or threatens time to market or plans of record for product/technology ramps. Generally, knowledge indicates there are no technology conditions that would restrict the deployment of this strategy. In fact, there is evidence that time to market, product quality at launch, and new product costs are all positively impacted by a successful execution of this strategy.

1.2.3.1 Disruptive Organizational and Technology Threats

In his book, *The Innovators Dilemma* (Christensen, 1997), the author discusses risks to companies introduced by disruptive technologies and innovations. He uses three findings to help characterize why good management can still lead to failures:

- balanced use of sustaining versus disruptive technologies
- technologies can outpace market needs
- an organization's investment strategies may be biased based on historic success, and do not take into account disruptive shifts in technology

For this book, disruptive threats (regardless of type) are defined as:

> products or services that can be seen as underperforming marketplace products but meet minimum customer needs at a better price/value point.

The author explores the potential that the underlying technology on which many of your products are based is advancing at a rate that exceeds the end market's needs. Based on this premise, those solutions that in the past

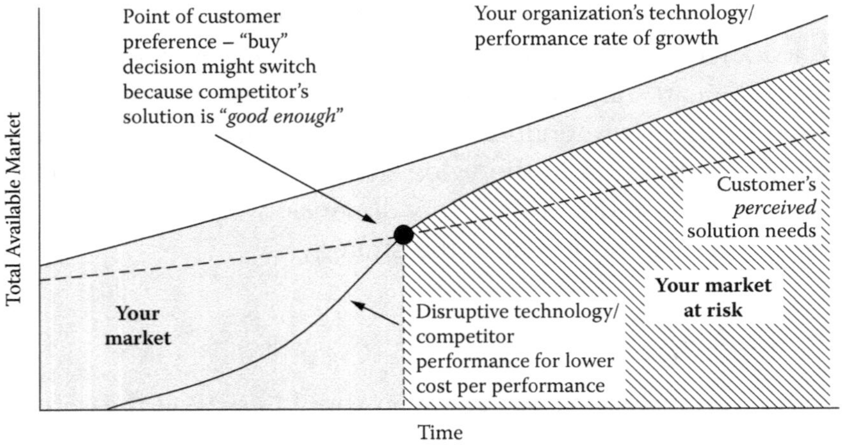

Figure 1.2 Disruptive technology's effect on an organization's market.

were deemed insufficient can intersect the customers' requirements entering the low end of the market, with a very competitive price, and performance that is "good enough." Based on the mismatch in the evolution of customer requirements and the evolution of technology, these solutions quickly improve, gaining more market share and eventually becoming the dominant force in that market.

In Figure 1.2, the presence of a disruptive technology (or competitor) is seen entering from the lower left at a lower price point. Because they are in the low end of the market and their rate of growth is much steeper than that of the traditional incumbent, they will continue to provide solutions across a broader range of customers' needs. Without the new entrant in the market, the existing organization's market is strong, stable, and growing. The moment the disruptive technology's capabilities exceed the customer's perceived solutions need, their lower price point, or other product attribute, it makes an attractive decision option.

Disruptive threats usually approach the customer's requirements from a very nontraditional manner. They typically display significant cost advantages or convenience factors.

How does your strategy stand up to this threat? What does an organization that is Learning with Lean do in the face of disruptive technologies? The change-hungry organization described as enabled by LWL will naturally be focused on challenges such as:

1. being responsive to the customer's needs, not overdesigning or supplying solutions,
2. greatly simplifying the supply chain, dramatically lowering your cost structures,
3. improving responsiveness to market changes including reduced cycle time for New Product Introduction (NPI), and
4. leveraging processes that yield only products or services of excellent quality.

This combination of LWL tools and practices can enable the organization to respond quickly and effectively to any change in customer dynamic, from product performance issues through supply chain demand cycles and, indeed, disruptive technologies. Relative to this threat, this strategy is providing a potential solution.

Based on this assessment of threats, the LWL strategy is deemed to be effective against most of the potential risks the organization will face. The recipe is not what to do, it is creating a culture to want to change and providing the tools and processes to use in the evolution of the organization into becoming a Lean Learning Organization.

1.3 Overcoming Organizational Inertia

> All that is necessary to break the spell of inertia and frustration is this: Act as if it were impossible to fail. That is the talisman, the formula, the command of right about face which turns us from failure to success.
>
> **Dorthea Brande**

The initial challenge in the transformation is all about overcoming organizational inertia. In the quote by Dorothy Brande, she sums it up by implying that you can overcome the frustration and inertia of the status quo by firmly believing that you cannot fail, and by using your courage to implement an about-face is the sure way to being able to move from failure to success.

In this section, concepts of organizational stress, both internally and externally, efforts required to create internal sense of urgency when no crisis exists, and the natural behavior in an organization to resist moving away from the status quo will be covered.

1.3.1 External Crisis Impacts to Inertia

A quick check of Webster's dictionary (Kellerman, 1981) for stress yields the following:

> Stress is a constraining, urging, or impelling physical force exerted.

In western culture, stress in our lives is considered a negative thing, but if examined more closely it can be seen that the negative implications of stress are the result of the prolonged exposure to high levels of stress, often higher than what the situation requires. In their book, *The Vitality Quotient* (Earle, Imrie, & Archbold, 1989), the authors offer the comparison between stress and the tachometer on your car. They postulate that you need some element of stress to get out of bed in the morning (i.e., a stimulant to change the current state). The negative aspects of stress occur when we do not use it or manage it properly, and we have our stress tachometer at 6000 RPM to make coffee in the morning.

With the definition of stress in place, it is easy to understand how organizations can respond with varying degrees of urgency and action depending on the nature of the external threat or stress. There are numerous examples of organizations taking drastic actions, or making dramatic changes in strategy when under a competitive threat that could jeopardize all, or part of, their organization.

An example is in Paul Carroll's book, *Big Blues: The Unmaking of IBM* (Carroll, 1993), where he decomposes the changes in competitive landscape for IBM yet their cultural attitudes of entitlement and market command prevented them from recognizing the global threat until it was almost too late.

Organizationally, the norm appears to have been established that the greater the crisis (stress) the more radical the solution will probably be. By communicating the nature and degree of the crisis, management injects the amount of stress required for people to accept the proposed changes. In many cases, the trauma to workers requires additional professional help to cope with the changes, especially as it relates to loss of jobs or reduced income or benefits.

Figure 1.3 shows the typical relationship of internal versus external stress inducers and their effect on altering, or affecting, the rate of stress experienced by an organization.

It appears counterintuitive that the greater the threat the less additional stress or energy is required by management to induce a change. The general

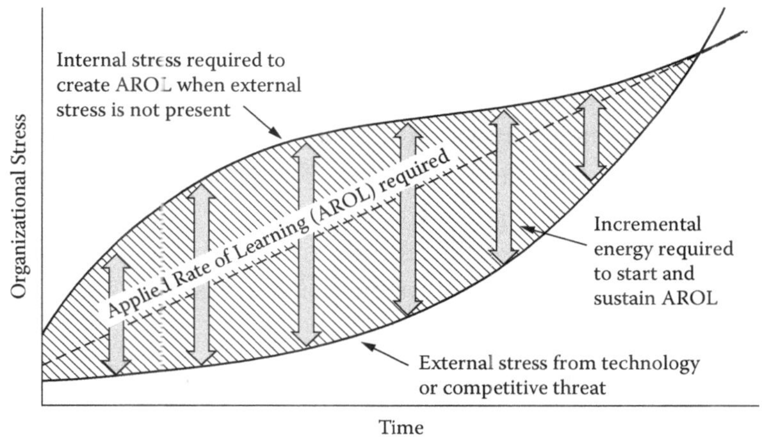

Figure 1.3 Internal versus external stress relationships to organizational change.

populace has sized up the situation and, often through fear of gossip, created an even bleaker picture of the future than what occurs. In this scenario, management is trying to control the rate and direction of the change energy in the group. Focus is critical and the "call to action" is not necessarily required other than to formalize the direction and quell fears, enabling the organization to start to move in the new direction. As much as nobody likes the drama, managing in a crisis is relatively easy.

Examples of stressors, internal or external, are:

- External stressors
  - new competitor in the market
  - hostile takeover or predatory practices by competitors
  - environmental catastrophes
  - significant regional or local events causing disruption of normal operation
  - major failure in one or more supply chain partners
- Internal stressors
  - poor quality of internal deliverables between functions
  - rework/redo loops
  - too little resources or time
  - poor/missing definition of requirements
  - no synchronization between functions
  - no cross-functional communication or feedback
- No or few standardized work methods

There is no doubt that the short lists of internal and external stressors fall short of what actually happens in the normal course of an organization's activities. If you look at the full range of stressors, you see they can create an environment of crisis for an organization—a crisis that adds urgency and focus to finding and fixing what is wrong. What happens when the stressors either are absent or exist at such a low level they are perceived as only noise? In the next section, we will talk about creating a sense of urgency when none exists through internal or external stressors.

1.3.2 Creating Internal Urgency with No External Crisis

In Figure 1.3, the greatest gap in the organization's ability to change occurs when the external threat is minimal and the need to change is early in the process. This is the area where most organizations will not see the need to change. Their ad-lib lines will be well rehearsed, explaining away any external indicator with historical references that may or may not be valid. In the case of no competitive threat, the necessary internal stress required to produce an increasing rate of change increases for a prolonged period.

- What happens if there is no competitive threat?
- How do you initiate change when you are gaining market share or winning in your field and successful?

In their book, *Changing for Good* (Prochaska, Norcross, & DiClemente, 1994), the authors describe a "transtheoretical model" that emphasizes how changes occur, how they become sustainable, and when results from changes can expect to be realized. In Figure 1.4, the lessons from their model are applied to organizational change and a model of deployment, change energy, and results emerges. In the model there is an "event horizon," seen as an identified spot, where the amount of energy required to sustain the rate of change drops continuously until it reaches a low investment, yet the rate of change continues to rise and results continue to be realized. This point is when a sustainable competitive advantage becomes evident and the changes have taken a permanent hold within the organization.

As a successful organization, you will not be able to rely on the external threat to provide the stress in the organization to change. Based on the graph in Figure 1.4, the amount of energy needed will be significant to climb the first "hill" of deployment and achieve the modest gains. This rate of energy will only continue to rise during the initial phase of your journey.

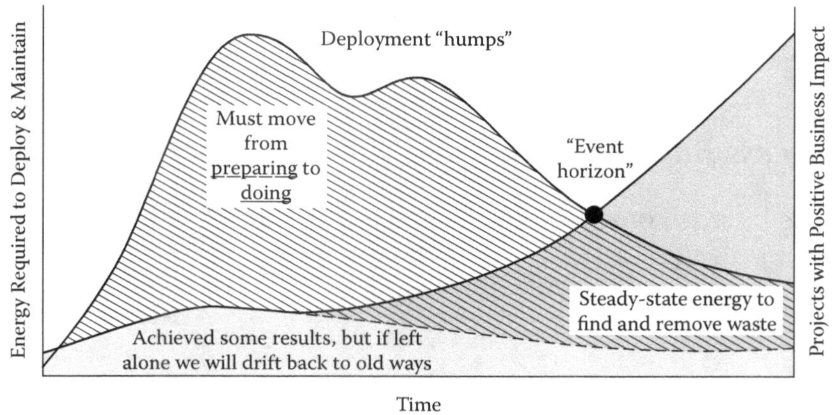

Figure 1.4 Deployment energy versus change in sustainable results.

Furthermore, just when you think that the organization has reached a critical mass and is sustaining, you will need to re-invest energy into the system to ensure that it actually does not slide back to the state of inertia from which it came.

This challenge of applying the necessary stress into the organization to achieve the rate of change required is the single largest obstacle to overcome. Your ability to deliver this level of stimulus over a prolonged period is a predicator to beginning this journey. In later chapters, various ways to assess your readiness to undertake this role will be explored.

1.4 Recognition of the Risk of Inaction

> There are risks and costs to a program of action. But they are far less than the long-range risks and costs of comfortable inaction.
>
> **John F. Kennedy**

It is easy to recognize that there are costs and investment required to initiate a transformation and sustain the change. Much like Kennedy's quote above, while there are risks and costs to move forward, if you stand still and be comfortable in the status quo, the long-range risk will overtake you. This section discusses how success builds the inertia that in some cases supports and in others contravenes the forward movement for transformation. Additionally, there will be exploration about how ignoring certain risks can stall the organization's forward movement. The take away from these discussions is that if you do nothing, performance will degrade over time as a natural result.

1.4.1 Success Builds Inertia

> Success breeds complacency. Complacency breeds failure. Only the paranoid survive.
>
> **Andrew S. Grove**

There are many examples of the phenomenon pointed out by Grove. In normal activities, there are numerous examples of successful organizations not responding to changes in the market until it is too late. Evidence of this phenomenon likely exists in your own successful organization. You may witness examples of "group think" —no one challenging the status quo, ever-increasing resources applied to the dominant market segment with little or no return on investment (ROI), and performance degradation in product development or time to market while still holding the dominant market share. Whatever forms this condition manifests itself in your organization you must recognize that the energy required to move the organization would be many times the amount of inertia that exists.

Note this is not a linear relationship. As seen previously, the excitation energy required to produce tangible benefits as seen in the rate at which the organization is learning is significantly higher than those needed to implement short-term gains. Many short-term gains can be accomplished by focus and management attention. This is symptomatic of "just work a little harder" approach to achieving this goal. Inertia from existing success compounds with the lack of external threat and becomes cumulative, requiring a greater amount of net energy, or stress, required to move the organization.

Figure 1.5 depicts a natural dichotomy and balancing act between two opposing behavioral norms, one that is required in all organizations. On the left side, it depicts control and discipline while on the other side it references reaction and response. An organization that is optimized has found an appropriate balancing point, where control and discipline must be encouraged yet tempered with the need to react and respond as the situation dictates.

As leader of the organization, this is typical of the dichotomy that you must handle during the transformation. In order to achieve the maximum

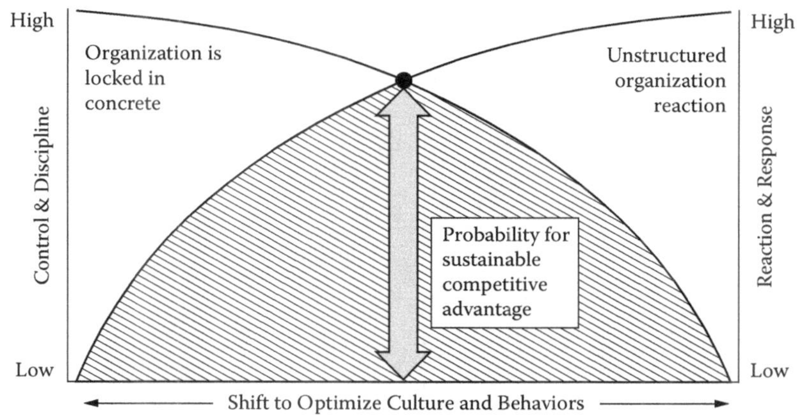

Figure 1.5 Dichotomy of optimizing control and discipline versus reactions and response.

probability for sustainable competitive advantage, you must seek to find the optimum point.

Depicted in Figure 1.5 is a declining curve, from high to low as it moves from left to right, representing control and discipline behaviors. A rising curve, from low to high as it moves from left to right, represents the behaviors of reaction and response. These curves, where they cross, provide the lowest point of resistance (similar to a eutectic point—lowest melting temperature—in certain binary compounds) and the best of both environments and maximize the probability of successfully establishing a culture that enables a sustainable competitive advantage.

This point, while evident in all organizations, is also unique in all organizations. It is driven by customers, suppliers, regulatory requirements, and internal performance expectations. The curves in Figure 1.5 are shown simplistically to represent the dichotomy of behaviors that must be balanced. Expect that the shapes of the curves are not symmetrical to each other, and they can take on different forms: exponential, "S" curves, linear, and so on, depending on the maturity of the organization and its environmental stresses.

1.4.2 Ignoring Risks May Stall the Organization

One final caution about a risk you will face. In many instances, the acceptance of change follows a distribution offered by author Everett Rogers in his book, *Diffusion of Innovations* (Rogers, 1995), with five categories of adopters during the introduction and deployment phases of a change. He

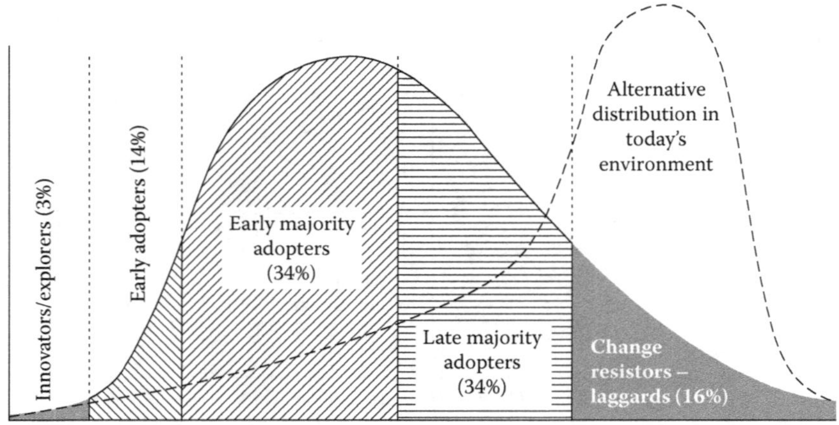

Figure 1.6 Change adopters distribution.

postulates that these five categories have a distribution approximating 3% for innovators/explorers, 14% for early adopters, 34% for early majority adopters, 34% for late majority, and 16% for laggards. Alternative arguments propose that the distribution is heavily shifted to the right with a longer tail to the left and a sharp drop off on the right. In Figure 1.6, in the case of LWL you will have early adopters, but there will be a number of factors that impede the rate of adoption.

Many people simply will not know what to do, how to apply the techniques of Lean, or challenge the status quo. The impact of what the early adopters accomplish may not be obvious to the extended organization. The resistance to this evolution will be high. Your vigilance and continued application of stress to the organization will be necessary to execute the LWL strategy. In the end, you will have to establish throughout the organization that the risk of not changing is greater than the risk of the change that you are proposing. In the successful organization, this is a significant hurdle.

1.5 Lean Enables a Learning Organization

> An organization's ability to learn, and translate that learning into action rapidly, is the ultimate competitive advantage.
>
> **Jack Welch**

As you drive into the use of Lean as a learning vehicle, you will be using OL as the approach to acquire new information and knowledge that results from many of the tools and techniques embedded in Lean practices. This will be leveraged using Lean principles to set a course of transformation where a desired end-state will be demonstrated behaviors of a Lean Learning Organization.

First, let us set baseline understandings for Organizational Learning, a Learning Organization, and Lean. Each of these is covered in this section. After these foundations, we will look at how and why Lean is one of the best learning vehicles to accelerate and sustain the transformation.

1.5.1 Organizational Learning

What is OL? OL is defined by Argyris with the concept that organizations learn because the individuals act on behalf of the organization (Argyris, 1977-55[5]):

> The individuals' learning activities, in turn, are facilitated or inhibited by an ecological system of factors that may be called an organizational learning system.

This organizational learning system is broken into six dimensions in Table 1.1. Each of these dimensions provides a view into the range of cultural dynamics exhibited in mature organizations that practice OL. It is important to recognize that as organizations mature they adopt greater levels of sophistication and integration into the fabric of who they are and how they act.

With an understanding of these six dimensions of OL, take the next step forward and use them to craft what your organization's unique view of the future looks like as it becomes a Lean Learning Organization.

1.5.2 Learning Organization

What is a Learning Organization? In his book, *The Fifth Discipline* (Senge, 1990), the author defines a Learning Organization as:

> An organization that enables learning of its members and continuously transforms itself using five main features: systems thinking, personal mastery, mental models, shared vision, and team learning.

Expanding the five main features noted in the definition (called dimensions in this text) can be found in Table 1.2. Each dimension is briefly described with the chief characteristic evident in an organization achieving

Table 1.1 Dimensions of Organizational Learning

Technical knowledge transfer	Knows the meaning of technical words and phrases, is able to read and make sense of data, and is able to act on the basis of generalizations; scientific knowledge is "propositional"; it takes the form of causal generalizations.
Narrative knowledge transfer	Organization stories enable participants to make sense of the difficulties and challenges they face; by listening to stories, members of organizations learn from each other's experiences, adapt the recipes used by others to address their own difficulties and problems.
Opportunistic application of knowledge	Often using materials or methods in unorthodox or unusual ways, and relies on resourceful and creative deployment skills and materials to meet each challenge in an original way.
Recipe knowledge transfer	The complexities of a manager's task frequently require use of a recipe, or an idea, that solved a particular problem in the past. Sometimes a manager may combine two different recipes or adapt an existing recipe with an important modification to meet a situation at hand.
Tacit knowledge transfer	Tacit knowledge consists of beliefs, ideals, values, schemata, and mental models that are deeply ingrained and are often taken for granted. This cognitive dimension of tacit knowledge shapes the way people perceive the world.
Systemic infrastructure for knowledge access and transfer	Methods and systems contained and controlled in an organization that enable fast, useable, repeatable knowledge access. Contained in the infrastructure are methods for knowledge capture, cataloging, referencing, accessing, and retrieval.

this level of maturity and performance. The premise is that these dimensions are essential to a mature organization striving to establish and maintain its competitive advantage, but are not enough to assure and sustain the advantage. Using a simple "necessary and sufficient" test finds missing elements of speed, customer voice, and resilience to disruptive technologies.

By understanding the dimensions of both OL and a Learning Organization, it enables you to advance into understand how Lean is used as a learning

Table 1.2 Dimensions of a Learning Organization

Systems thinking	A set of habits or practices within a framework where component parts of a system can best be understood in the context of relationships with each other and with other systems, rather than in isolation.
Personal mastery	Commitment by an individual to the process of learning.
Mental models	Assumptions held by individuals and organizations.
Building shared vision	Shared vision is important as it creates a common identity that provides focus and energy for learning.
Team learning	Learning Organizations have structures that facilitate individual and team learning with features such as boundary crossing and openness.

vehicle. Until now, improvement activities have been talked about obliquely in both sets of dimensions. As you move toward the next assessment fostering LWL, all of the questions and topics will be specifically focused on Lean philosophies, behaviors, and actions, those that systematically identify improvement opportunities and use discipline approaches for the tools to be able to find a root cause and preventative solutions for implementation.

1.5.3 Lean as a Learning Vehicle

The real question is how can you build the lowest cost environment (processes and systems) that gives the highest tipping point value to the organization? Tipping points are the benefits—tangible and intangible—that enable success in the market and competitive strength to resist organizational threats.

What is Lean? The traditional definition is:

> Lean is a practice that considers the use of resources to achieve any goal other than the creation of value for the customer to be wasteful, and thus a target for elimination.

Let us restate it in even simpler terms that encompass the premise of using Lean for the implementation of OL:

> Lean leadership is process minimalism in belief and execution—only actions that transform resources into customer value can remain.

We can break down the concept of process minimalism into its components—process and minimalism. "Process" is well described and understood in other readings (Camp, 1989; Cobb, 2003; Rother & Shook, 2003) and does not require additional exploration in this discussion. The concept of minimalism can best be defined:

> Minimalism is the focus of work set to expose the essence, or essentials, of an action through eliminating all non-essential forms, features, or concepts.

Why focus on Lean minimalism? Isn't Lean and its focus good enough? The mechanics of Lean can often become one of your own weaknesses because the steps become mechanical and Lean opportunities will be overlooked unless the concept of minimalism is included to ensure the challenges as to what is really needed.

Often, it is not about whether a customer will pay for the steps—many times they would not be willing to pay for certain actions of an organization, but they do see value in the changes to deliverables the actions provide. This allows you to weave into the overall theme the concepts of "soft" deliverables that provide "customer delight"—a crucial aspect to a sustainable competitive presence in your market.

Simply stated, the benefits of Lean minimalism are:

- **Speed of execution**—fewer steps to execute to provide the genuine content expected by the customer
- **Quality of results**—fewer touch points provides a natural increase in quality of output
- **Ease of learning**—simple solutions flow between processes, OL increases because language and methods become common
- **Agility for change to meet new expectations**—with fewer steps, simpler language, and increased quality of output, the organization is more responsive and likely to change to shifts in market and customer needs
- **Intelligent automation**—minimalized process construction makes automation easier to define (fewer variations, work around), simpler to execute (fewer steps), and produces less waste through consolidated inputs and outputs

Table 1.3 Lean Traits that Enable Organizational Learning

Simple to use	Easy tools, methods, and analysis; frequently "back-of-the-napkin" discussions are all that is required.
Useable by everyone	Applies from CEO/Senior Executive to individual contributors; standard work and Management Standard Work (MSW) are seen at all levels.
Applies to all types of work and deliverables	Every type of task or action can use basic Lean tools and methods to find and eliminate waste.
Driven through data	Data is the backbone of Lean improvement; it provides the foundation to analyze and find hidden waste.
Consistent over time	Few changes in tools/methods construction; they have been used consistently in a wide variety of challenges.
Enables reinvention of work and skills	Lean methods provide a structure for individuals and leaders to re-craft their work, eliminate non-value added tasks, and enable reuse of recovered time for higher value-added skills building in all employees.

How does Lean enable OL? Six traits of Lean enable the core tenets of OL. In Table 1.3, the traits are mapped to how Lean enables OL and the simplicity of how it is used.

To help explore these six traits of Lean and their linkage to the core tenets of OL, more discussion follows.

1. **Simple to use**—The tools, methods, and data analysis used in Lean are exceedingly simple. Fundamental tools are used, such as Pareto charts, run charts, and data segmentation. Individuals and teams can use very simple statistical tools (embedded in most software spreadsheet applications) to quickly calculate how capable a process is to meet/exceed customer expectations.
2. **Useable by everyone**—Every single member of an organization can (and should) use the simple methods of Lean to find hidden waste, remove it, and ultimately optimize a specific task. While the proportion of time and types of activities spent by individuals varies based on their levels of responsibility, everyone is accountable to demonstrate how they know and use Lean to drive them toward being a Learning Organization.

3. **Applies to all types of work and deliverables**—Lean methods are directly applicable to any type of work, task, or project. Using Lean to help drive the concepts of being a Learning Organization has demonstrated its impact in private and public organizations, nonprofit and profit, manufacturing, service, government, healthcare, and education alike. The only difference is which methods are used, in what sequence they are used, and how the outcomes are applied within organizational constraints.
4. **Driven through data**—This is simple; data is king. The cornerstone of good practices for identification of waste, and improvement and transformation to sustainable learning, is use of data to tell you what is real and what is not real. Analysis tools are usually simple (some more complex issues require more sophisticated tools, but these are not as common as expected), the findings are easy to interpret, and the outputs lead to simple solutions.
5. **Consistent over time**—Lean and how it enables transformation toward a Learning Organization is evolutionary by nature. The methods and tools used have been around for many years. Little real innovation in tools has occurred, only new ways to think and use the tools and methods. The purpose of this book is to show a methodical approach to reinventing your organization's approach to excelling in performance excellence.
6. **Enables reinvention of work and skills**—By its very nature Lean thinking and methods cause work tasks to be streamlined and reinvented. When enough tasks have been streamlined (wastes removed), the employee responsible for the set of tasks will be repurposed to higher value-add activities. Those higher value-add tasks often require new, more complex skills and competencies. The result is an employee with a great overall skill set, someone adept at identifying and removing waste, and a person with a more robust and adaptive approach to being successful.

Let us take a moment to explore how Lean is flexible and can be used to examine and improve the individual processes. Its power is that the same language, tools, and methods are used to expand the process analysis into system analysis. Frequently, the improvement effort on a small process exposes opportunities in the larger system it fits into. Many of the dependencies of the larger system are based on other processes that feed into it. If there are "waste-laden" inputs (e.g., input errors, difficult to use inputs, poorly timed inputs, inputs requiring rework/redo actions before being useable), the overall effectiveness and efficiency of the larger process is

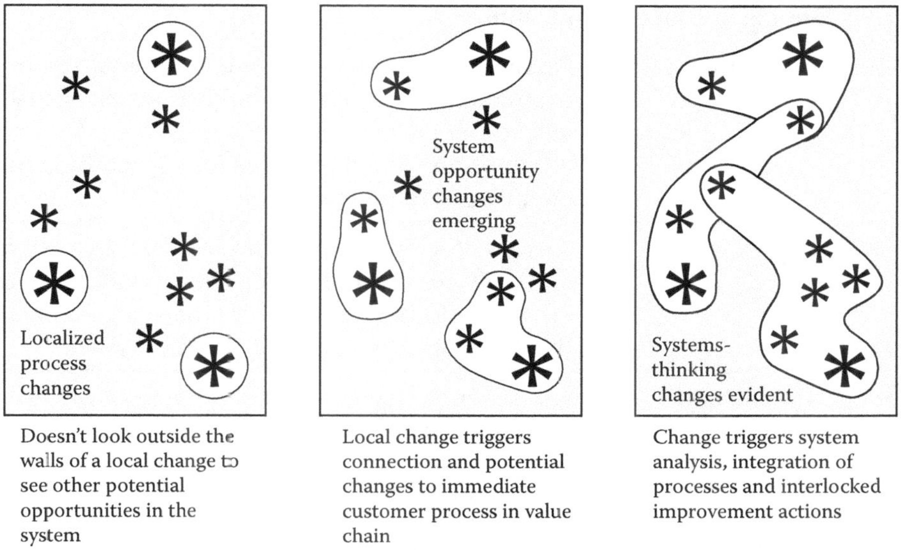

Figure 1.7 Synergy of improvement actions with AROL maturity.

impacted. It is this transition from fixing small process steps at an individual contributor's level to choreographing the improvement of the larger subsystem by a systemic application of targeted improvements and the individual contributors level that marks the organization evolution in its application of Lean. Just as often, the outputs of the improved process feed receiving processes that are not optimized for the faster, cleaner, higher quality inputs. Frequently, fixing a process leads to fixing a system and the system integrates with other systems.

In Figure 1.7, the uncovering of linkages between improvement opportunities, or improvement actions, is a wonderful by-product of the power, flexibility, and usability of Lean to find and learn what is needed from an organization to better succeed for its customers and in its market.

Using Lean just as a toolkit, or series of methods, is only starting the journey and will not uncover the hidden power it brings to organizations that embrace it as part of its culture. In Table 1.4, the dimensions of a Lean culture are explored. Each of these dimensions has "tattle-tales" visible and evident in daily operation of the organization.

Why worry about the organization's culture? Let us recognize that the organization is already successful, and it got there in part through its employee's embodiment of positive cultural behaviors. The premise of this book is about elevating the organization above the current level of

Table 1.4 Dimensions of a Lean Culture

Uses One-Touch Flow (OTF)	Pervasive approach for process minimalism. In examining every process, large and small, seeking to find the fewest possible actions and every action, it transforms the deliverable into something that is meaningful to the customer.
Requires standard work	System efficiency is maximized when consistent methods are used to do repetitive work, regardless of who is doing the work, or where it is being done; this includes Management Standard Work.
Intolerant of waste and bureaucracy	Consistent behaviors that challenge all processes as potentially wasteful (the entire process, not just the incremental steps). There is a natural discomfort with accepting the status quo.
Focuses on error-proof design	Processes and designs are architected and built to prevent errors from occurring or, if they occur, from getting to the customer or interfering with effective and efficient execution of partner processes.
Knows its core processes	A core process is at the heart of an organization's ability to meet its mission and achieve its desired value-add to its customers. All core processes must be identified, performance measured, and results validated with the Voice of the Customer (VoC).
Improves what is important	Using core processes (which target mission success for the organization) and their delivered performance, gaps should be identified, normalized, then "racked and stacked" using a prioritization schema integrating importance, impact, speed of improvement, and investment.
Focuses using the Voice of the Customer (VoC)	VoC is used as a focusing agent when potential improvements have to be prioritized and action plans implemented. A common misuse of the VoC is to use it as the reason for change (it should not be the catalyst).
Treats its supply chain just like itself	An effective Lean organization reduces its supply base to levels consistent with the velocity of its customers' consumption and resilience to world events. Integration of the supply base as an extension of the organization is crucial to aligning capacity and minimizing supply constraints.

Table 1.5 Four Dimensions of Knowledge Lifecycle

Creating (or acquiring) knowledge	An individual or group activity that is normally small in scale and isolated and uses the methods and language of the subject matter experts.
Capturing individual learning	Capturing knowledge and experience takes the form of publications, after-action reports, post-mortems or lessons learned, interviews and discussions, and presentations of all sorts. This must include organizing knowledge so people can find it (e.g., multiple structures/hierarchies for rapid searches) and safe storage to ensure that it will be available in the future.
Transferring knowledge	Transferring knowledge requires ready searchability via browser-activated search engines; content retrieval is also needed, which requires a communication and network infrastructure. Tacit knowledge may be shared through communities of practice or consulting experts.
Mobilizing knowledge	Involves integrating and using relevant knowledge from many, often diverse, sources to solve a problem or address an issue. Integration requires interoperability standards among various repositories. Using knowledge may be through simple reuse of existing solutions that have worked previously.

performance into being a Learning Organization that uses Lean to help it create and sustain a competitive advantage.

It is critical to map the changes in organizational hunger (desire, willingness, and capability) to offset threats that are emerging as real or possible (technology, commercial, or competitive). To be sure, this concept applies to all kinds of organizations—profit/nonprofit, private/public ownership, and public sector at any level.

At this point, it is important to mention the dimensions of the knowledge lifecycle. Crucial to the success and advancement of OL, a cornerstone is data access and visibility, plus rapid access and use of information and knowledge. In Table 1.5, the dimensions are creating or acquiring knowledge, capturing individual learning, transferring knowledge, and mobilizing knowledge. Each of these is described in detail, defining what makes up the dimension, its impact, and its use.

The importance of data and information needs to be emphasized in the overall transformation to a Lean Learning Organization. Data availability and use is a central hinge factor in the success or failure of what you are about to start. Everyone has heard the old euphemism "data doesn't lie, but you can lie with data"; never has something been so true.

You will need to spend time on looking at data purity, availability, reliability, analysis, and translation into information and knowledge. Also, know this does not mean that you need huge data systems, IT architectures, or large investments in automation. On the contrary, the simpler the system, the less data required to determine if it is in control and how capable it is to meet customer expectations.

1.6 Transformation Is an Endless Journey

> Focus on the journey, not the destination. Joy is found not in finishing an activity but in doing it.
>
> **Greg Anderson**

There are many discussions about undertaking a journey, but none is as critical as recognizing it must allow for a wide berth of how the steps follow each other and recognition that there will be interruptions, deviations, and, in some cases, wandering, due to internal and external pressures. The leader must be able to know they are coming, see them as they arrive, and guide the organization through them while keeping true to the vision and course.

This section discusses setting of the journey, staying the course, and examining your personal strengths and opportunities so that you can better prepare yourself to lead the transformation.

Assessment, as described using the premise of this book, takes on examinations of the leaders and the organization as a whole. It uses a range of instruments to determine strengths and challenges for specific behaviors, methods, and skills, plus tools that inventory personal resources for influence and change by the leader, and organizational performance as an indexed measure. These include:

- Leading a Learning Organization
- Lean Leadership
- Lean Thinking and Acting
- Credibility Cash Index (CCI)
- Organizational Performance Quotient

Each of these is supported with descriptions and discussion later in this chapter, plus examples are included and full exhibits follow in the Appendix at the end of this book. The Appendix provides a complete view of the various assessment instruments and information of how to interpret the findings from each one.

1.6.1 Phases in the Transformation Journey

> Culture does not change because we desire to change it. Culture changes when the organization is transformed; the culture reflects the realities of people working together every day.
>
> **Frances Hesselbein**

Organizational transformation really is a simple concept, but is loaded with many challenges and difficulties to conquer. It is easy to recognize that it is a journey over time; one that, by design, should be continuous in nature and feeding.

In Figure 1.8, the actions taken during the transformation are represented as a looping system. These steps are not new or novel in their construction

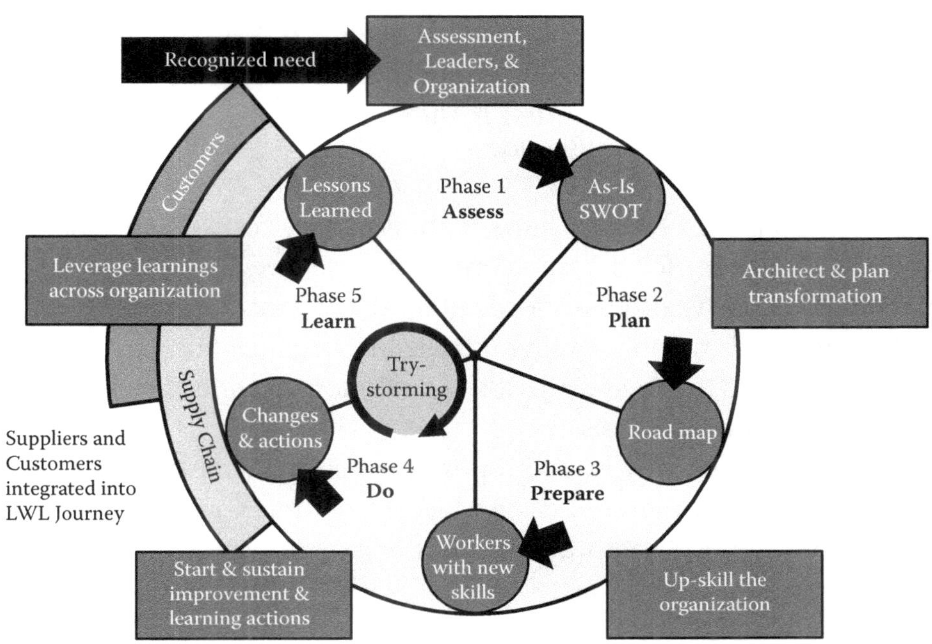

Figure 1.8 Five phases, transformational model.

or arrangement, but the simplicity of the looping system brings to light the five simple steps needed to understand current state, design a forward direction, start changing, and then capture and leverage learnings.

Everything starts with an assessment of the "as-is" for the leadership team and the organization as a whole. Outputs are strengths and gaps identified. Everything starts with an assessment of the "as-is" for the leadership team and the organization as a whole. Outputs are strengths and gaps identified. Since we are dealing with a cultural change much of the assessment will be focused on the adaptability and demonstrated learning ability of the organization and individuals. From this current state you can begin to see where you have capabilities that can be leveraged…these will become the girders and foundations of the bridge or transformation elements that need to be put in place to move the organization.

Embedded between Phase 4 and Phase 5 is a small loop, Try-storming. This is the micro-learning cycle that happens when small improvement efforts are started (typically improvements that happen at the individual or small team level). This is very similar to rapid prototyping in software development environments.

Briefly, this is defined as:

> the integration of brainstorming and simulation techniques combined with rapid construction of a prototype solution (out of anything that is available) allowing people to visualize, touch, and further improve on an initial idea.

In this section, the first step starts with assessing yourself and the organization. As the leader for the transformation, you need to know your personal strengths and weaknesses for leading the organization in the areas of Lean Thinking and Action, Leading a Learning Organization, and CCI. Each of these is explored, tools are provided, and interpretations are given for various outputs from the tools.

Throughout this book you will repeatedly see the simple five-phase cycle of "assess, plan, prepare, do, and learn" with various activities and actions assigned to each with deliverables exiting and moving from phase to phase.

It is also stressed that the cycle is not a one-time excursion. A true Learning Organization will have this built into its fabric, or DNA, as part of the way it learns and perpetually executes.

1.6.2 Phase 1: Assess

This section is all about assessments at the leader and organization levels; for both Lean and Learning Organization, strengths and weaknesses are the first step. In Table 1.6, there are five foundational tools used. Uniquely for the leader there is the evaluation of how much credibility he or she has to effect change. This is called CCI. Each of these instruments will be explained later in supporting sections, but recognize that it is a sound practice always to take the first step by figuring out where you are in the journey, what strengths you already have, and what significant opportunities are in front of you.

You, and the organization, will come back to this step many times. The improvement loop is cyclical and the speed at which you go around the loop is an indicator—a speedometer if you will—on the rate of change the organization is experiencing.

Each of these tools is described in more detail in the Appendix at the end of this book.

The construction of the assessments is focused on finding gaps in knowledge of and capability to use the tools, methods, and culture required to make the transformation. As they say, you only get out what you put in. There is a need to be brutally honest and find out exactly what is going on in the organization.

During the assessments at both the leader and organizational level, there is a need to clearly identify the stressors in the organization. These need to be both internal and external. The gaps can be identified in these findings and used to help build the bridge to an architecture that is properly sized and focused on changing the right things.

1.6.3 Phase 2: Plan

Now that you have painted the organization's landscape of its strengths, weaknesses, and gaps you must quickly synthesize an architecture and plan for what needs to be changed and when it needs to be changed. Usual considerations are time, cost, and difficulty to put in place the changes, some of which could be foundational to the organization, especially when you are talking about ones that impact the culture and behaviors.

In the typical sense, architecture for change mirrors an architect's plan for building a structure. It is composed of a visual rendering of the final structure, supported with a floor plan that defines how the structure is laid out

Table 1.6 Assessment Instruments

	Instrument	*Used by*	*How it Works*	*When to Use*	*Using the Results*
Learning Organization	Leading a Learning Organization assessment	Leaders, all levels	10 questions, answered	Kickoff, then biannual	Part of developing development plans, personal and organizational
Lean Leadership	Lean Leadership assessment	Leaders, all levels	38 questions, answered	Kickoff, then biannual	Part of developing personal development plans
Lean Thinking and Acting	Lean Thinking and Acting assessment	Everyone	24 questions, answered, responses collected across organization	Kickoff, then annual	Part of developing and managing personal and organizational development plans
Credibility Cash	Credibility Cash Index 360° inventory	Senior leaders	5 questions, answered by self, manager, peers, subordinates	Semiannual	Used frequently to adjust working styles and networks to influence change methods
Organizational Performance	Organizational Performance Quotient inventory	Leaders, all levels	5 vitality traits evaluated using organization-specific indexed measures	Kickoff, then annual	Part of routine organizational system performance management methods

with interconnecting spaces for key activities (e.g., meeting, social gathering, work areas, and so on), specifications of the materials of the structure (both type and quantity required), and finally construction instructions to prepare the building site and assemble the structure.

The following analogy draws parallels to the architecture of a building and what you must design and provide for a well-planned and articulated roadmap for your organization's transformation journey.

- **Rendering of structure**—An architect creates visual renderings of what the finished structure looks like; in organizations, the leader creates the context and mental models of the endgame, defining what the finished state of the organization should look like.
- **Floor plan for key activities**—An architect designs a floor plan for effective/creative space utilization; the organization leader defines the changes to culture and behaviors that are expected as the transformation unfolds. This also includes potential organizational structures, changes to roles and responsibilities, and shifts in decision-making to lower levels of the organization.
- **Specifications of materials**—An architect will specify the materials used in construction, while the senior leaders of an organization will specify the tools, methods, and systems to be installed and used during and after the transformation. These will be specific and consistent across the organization so that there is essentially only one transformation executed.
- **Construction instructions**—Architects create specific assembly instructions that use the specified materials exactly to produce a structure mirroring the original rendering. Organization management is accountable to supply specific instructions that interlace the tools, methods, and systems into a systemic operating model that increases organizational performance. This includes expectations of senior leaders and managers regarding how they will review and use methods that establish standard work (either worker or management) as the norm.

In the beginning, you need to have a vision of the end game—a rendering of what the target looks like. That is only the starting point. It is common to think that the end state is really the "end state," but that is where you start—with the end in mind. From that picture you can step forward, articulating actions required to move from plan into deployment (moving from Planning phase to the Prepare phase).

One of the first actions is to be very specific on the central philosophy and practices you will use to effect the transformation. There are many choices that you can make when selecting the core philosophical approach to delivering quality systems and or performance excellence. When we compared some of these methodologies (see Table 3.3) Lean quickly percolated to the top of the group as the vehicle to use. Supporting this position of philosophy and practices will be methods used to bring the knowledge and skills to everyone across the organization.

1.6.4 Phase 3: Prepare

Preparing the organization for transformation is about deployment and motivation. This is where investment in training, tools, and methods occurs. Leadership buy-in and visibility is crucial to making sure the organization sees evidence "at the top" and that this is not an "initiative of the month."

In this step, the organization is exposed to and expected to start using the methods and tools defined from the architecture. It does not matter exactly which methods or tools are prescribed; what does matter is that the entire organization is expected to use them and act consistently. Leaders must start demonstrating that they know and can use the tools—they are the beacons for change in their organizations and teams.

Organizationally at the top, the senior leaders and executives must be able to easily demonstrate their knowledge and willingness to use the knowledge during project and organizational performance reviews. Their ability to quickly ask the right questions, use the right language, and set clear expectations for the kind of root cause analysis and closed-loop corrective actions is paramount.

1.6.5 Phase 4: Do

Yes, real work needs to happen. Some parts of the organization will be fast learners. They will implement changes quickly and will be the teams, or work units, which provide early learning examples and quick process improvement wins that you can use to accelerate deployment to the rest of the organization. Using these teams or work units as role models accelerates less aggressive work teams toward becoming aligned, and then starting to pull together for the same objectives and goals, and starting to use and practice the methods.

In his book, *Good to Great* (Collins, 2001), the author provides an analogy of a bus with a driver, passengers, and many different seats on which the passengers can sit. Ultimately, this is the point where the driver must know the destination and the route, all the passengers must be on board with a desire to get to the same destination, and each passenger must be in the correct seat. Anybody who is not on the bus is going to be left standing at the curb wondering when the next bus will come, but will quickly realize that there is only one bus and it left without him or her.

1.6.6 Phase 5: Learn

Convergence of planning, preparation and improvement experience turns into organizational learning in Phase 5. It is where leaders and workers must evaluate the activities and impacts of improvement on their organization for "value add" contribution. Throughout the set of phases it is emphasized that you never transform just for transformation's sake. There must be a reason, a compelling reason for the transformation. In Phase 5 there are several critical activities that surface:

- Try-storming emerges as an essential tool. It promotes rapid cycles of solutioning that result in fast, cost effective waste reduction, delivers increased results for invested energy and helps promote an organization's learning at ever-faster rates.
- Use 'transformation tattletales' as simple, readily evident changes in behaviors that gauge the rate of learning and adoption of Lean practices.
- Not all work being done today should be done by your organization. Segment the work using critical review and find out who (your organization, your customers or your suppliers) should really being doing the work. Then, systematically move appropriate work upstream and downstream to your supply chain and customers.
- Benchmarking is compares current performance versus your competitors, or against the market. It requires focus and energy to be impactful — use the approach when necessary to set the right meaningful goals.
- In order to keep organization fear in check, reinvention of work and job skills is a must. Any approach streamlining must be done openly, with clearly stated purpose and a definition how liberated resources will be refocused/reused.

1.7 Measuring Impact and Rate of Learning

> If the rate of change on the outside exceeds the rate of change on the inside, the end is near.
>
> **Jack Welch**

The quote from Jack Welch summarizes well the essence of measuring the rate of change, and the organizational impact, based on the rate of learning. If you cannot change internally faster than the outside is changing around you, you are doomed to fail. Effective measurements of your current performance and trends are your yardsticks of progress, success, and competitive advantage.

This section explores moving the organization closer to becoming a Change-Hungry Organization and what that means. Measurement will be instrumental in accelerating the rate of change an organization is capable of making. Later, in Chapter 3, additional discussion will focus on specific measurement approaches as part of designing the architecture and planning for the organizational transformation.

1.7.1 Change-Hungry Organization Maturity Levels

With any large transformation inside an organization, the landscape for introduction, adoption, and use of new methods, tools, and culture is widely varied. There are examples of this maturity distribution in many industries and organization types.

Constructive challenge is a powerful tool. Why should you take the time to consider these methods, what they have to offer, and how your organization might be able to use then when you are already successful?

- What is a Change-Hungry Organization?
- What is different about it?
- Why are the differences important?
- Is this "pie-in-the-sky" or can I actually be part of one?
- Is leading one dramatically different from other types of organizations?
- Is the investment worthwhile?
- Do I have the commitment and credibility to make one happen?

These are all questions that come to mind when you hear about something like a Change-Hungry Organization and rightly so. A lot of books, lecturers, organization schools, and consultants all have their own spin on increasing an organization's performance to levels that are Best-in-Class.

The definition of a Change-Hungry Organization is:

> A change-hungry organization is focused on growing its sustained competitive advantage through a culture of rapid cycles of organizational learning and process minimization using one-touch flow and elimination of process waste.

In Figure 1.9, you can see five levels of maturity above zero (Resistance and Inertia) that progress from a simple break/fix approach up to growing levels of learning cycles to the highest level of a genuine change-hungry culture. An example of this is in the software development world using an approach called CMMI (Capability Maturity Model Integrated). It has very similar levels of maturity and growth to get from one level to another and contains specific practices and expectations. In their book, *CMMI, Guidelines for Process Integration and Product Improvement*, the authors describe how each level has specific expectations and performance traits required. In order to move from one level to the next higher one, new capabilities and practices must be put in place and demonstrated.

The approach used throughout this book parallels the same concepts as it drives to a transformation into a Change-Hungry Organization driven to command a sustainable competitive advantage. In coming chapters, the same levels of progression, including new skills and capabilities, will become part of the architecture, deployment plan, and its rollout.

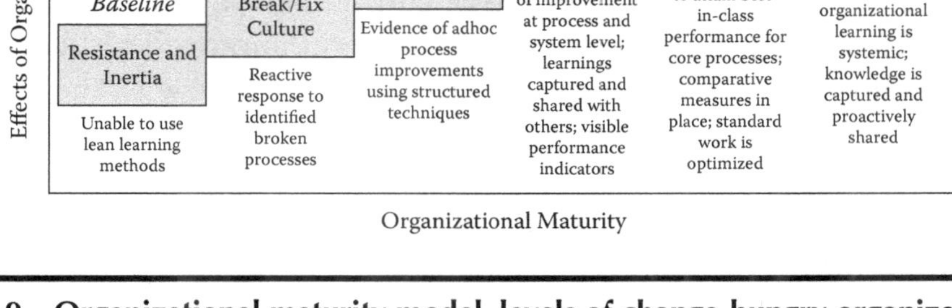

Figure 1.9 Organizational maturity model, levels of change-hungry organizations.

In order to move from level to level there are expected behaviors, disciplines, and actions as you move from Baseline (Resistance and Inertia) through more complex and mature levels and cycles of learning, all the way to Level 5 where the organization is demonstrating that it has and maintains a sustainable competitive advantage. At each successive level, there are associated improvements in operational performance and customer satisfaction, and increases in effectiveness, efficiency, and consistency.

Let us take a moment to explore each of the levels to expand on its general expectations and evidence.

Baseline—Resistance and Inertia: Organizations show signs of active push-back (verbal and process rejection of organizational learning characteristics), they are locked into the status quo, improvement actions are met with reactive emotion and a demonstrated inability to learn and use Lean as a vehicle for improvement. There is evidence that progressive examination of results and necessary changes to improve are ignored.

Level 1—Break/Fix Culture: Typical tendencies of the organization at this level are chaotic actions that only fix things that are not working; improvement efforts are mostly containment of process escapes and remediation of negative impact to customers. Frequently, measures of success only focus on the size of the impact, frequency of impacts, and cost implications to "make good" to the customer. There are few or no prevention efforts evident.

Level 2—Early Learning Cycles: Improvement actions at this level are usually anecdotal and based on pockets of passionate employees striving to increase results. Often the improvements are on a very specific process and there is little or no organizational learning. Inertia of normal operations inhibits systemic learning and use of process improvement knowledge and experiences. There is little or no system improvement evident.

Level 3—Sustained Cycles of Learning: Systemic improvement actions can be seen in many areas of the organization. While there are pockets of excellence that help create knowledge and improved results, cycles of learning can be seen at the process and system levels. There are visible measures of process performance at all core processes and most supporting processes. The measures show evidence of driving closed-loop corrective actions. Learnings are captured and proactively shared between work teams and units with the intent to use the learnings for faster implementation of improvements in other areas.

Level 4—Differentiated Performance: Processes and systems are being measured and improved proactively using analytical methods that characterize performance in terms of capability margins to requirements and show evidence of being compared to the best-in-class, best-in-task, and world-class benchmarks. There is evidence that competitive intelligence is used to determine SWOT against critical mission deliverables. There are standard workflows evident from the front-line worker level up to senior and executive management.

Level 5—Sustainable Competitive Advantage: Culture is present that expects and recognizes workers for aggressively finding and removing all forms of process and system waste across the entire organization. Evidence is present that cycles of learning and improvement are a daily way of life. The focus of the organization is on execution of processes that achieve the mission versus focusing on and optimizing individual process performance. Knowledge capture and reuse is rampant in all parts of the organization and there are systematic structures that enable ease of capture, reality cross-reference, and simplicity of reuse.

In the beginning of an organization's transformation journey, the many parts of the organization are at different levels of maturity, ability to change, and motivation to change. Some are very high in knowledge and willingness to change; others are locked in their ways and methods. They are resistant to change, but when they sense there is no choice, they acquiesce and start unfreezing their ways and methods. Others are steadfast in their belief that if they resist long enough, the need for change will pass and they will not have to modify the way they are doing work today. They will have weathered yet another initiative of the month. As a leader, this is perhaps the most tiring task of transformation—one that will take your conviction, courage, and resilience to lead.

1.8 Communicating to the Organization

The two words "information" and "communication" are often used interchangeably, but they signify quite different things. Information is giving out; communication is getting through.

Sydney J. Harris

As leader, one of your most critical deliverables to the organization is communication. Your teams should already be accustomed to hearing from you on your assessment of current performance, along with near-term and long-term objectives and strategies. In creating a Lean Learning Organization, your requirements to communicate at all levels in the organization will increase dramatically. The potential impact from this transformation is very large and yet your initial requests of the organization will be small. The fact that you cannot "buy" the solution but rather have to build it through your employees will be foreign. These are the first examples of many dichotomies that you will face along this journey.

Part of the problem arises from the fact that it is a journey. The end is not definable because it will continuously evolve based on the needs of the organization. The effort and energy that you ask the team to expend will be modest as they learn the tools and apply them to simple problems. The compounded impact of the entire organization using these tools with a passion to learn will stagger even the most ardent believer. It is with this backdrop that you need to structure your communication plan.

The term "porpoising" describes the transition from macro concepts to micro details and back. When you think of the needs of your audience, there will be elements for both macro and micro level input. This oscillation will cause angst for many who "simply want a straight answer." The concepts will be foreign to many, and you will face many skeptical listeners. Your job is to lead with the resolve and conviction necessary to ensure that the extended organization understands that you are embarking on a journey that is neither optional nor selective as far as their participation is concerned.

The dichotomy in focus is readily visible in many organizations—the need for attention to detail in the execution phase versus the need to be directionally correct on strategy. Why does this normal condition cause more resistance in the LWL initiative? Most organizations are used to accepting input on the strategic initiatives in terms of market segments, technology, product roadmaps, or evolutions in service offerings. In other words, they are expecting the future to be defined in terms of their current portfolio of products and services, or by items that are adjacent to them. Since few organizations identify themselves in terms of their processes, it is difficult to frame the future state or strategy in terms that resonate with the full employee base.

This represents the crux of the problem. Extrapolating a future state powered by the strength of LWL will be difficult for the organization to easily grasp. In any communications with the team, you must first address their needs to have both the macro- and microelements reiterated and reinforced. You have to be ready with your ad-lib lines well prepared, and indexed for the right level of macro and micro reference for each opportunity. No matter how well prepared you are, recognize that this is essential to assurance and reinforcement of the direction and recognition of the effort required by the organization.

1.8.1 Expanded Need for Different Types of Communication

In this early stage of moving an organization, which is already performing well, to higher levels of performance targeted on sustained competitive advantages, precise target communication is a critical asset.

A quick list (certainly not exhaustive) of types of communication venues includes:

- Large group meetings/discussions
- Town halls
- Skip level
- Stand-up meetings
- Brown bags
- Intranet and internal electronic communication forums and vehicles
- Organization newsletters and news flashes (regardless of media)

Not to be overlooked, the most impactful level of communication is the demonstration of expected behaviors, or behavioral changes, by the leaders. This is where "walking the talk" is best exhibited and provides the greatest yield in communication energy.

1.8.2 Dichotomy of Conversation

If different types of communication are critical, it makes sense that different levels (content-specific) are required based on when, where, and to whom the communication is being held. Think about the difference of talking with a senior leadership team versus talking with an individual contributor on the shop floor or in a work area. In Figure 1.10, a representation of the dichotomy of conversation content shows that early in the transformation cycle there is a need for a great communication focus (quantity, duration, and type) on messages at the broad horizon level.

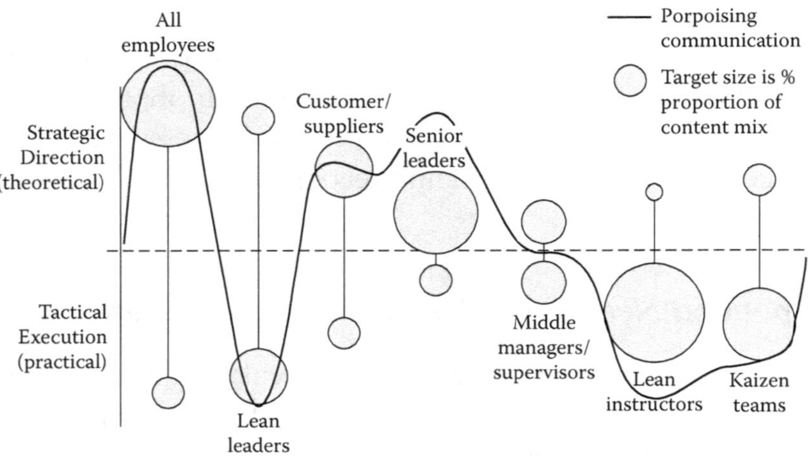

Figure 1.10 Communication porpoising, content proportion by audience segment.

Using a targeting metaphor, these messages are the ones that describe which barn you should paint a target on. They help set the direction for the transformation, acknowledge the highs and lows that will be encountered, and enable senior leadership the ability to express their passion, commitment, and expectations of the organization. Conversely, there are fewer messages that have very pointed specific or directive content. The organization is not ready for a large volume of these kinds of messages yet. These messages should be designed to get the ball rolling, and then adjust the level and type.

As the organization begins to gain momentum, the messages are refined to better describe what type and size of a bull's-eye target you should paint on the selected barn. Messages begin to shift from broad to more specific. Directives set crisp expectations, goals, and behaviors. Dedicated teams are formed. Measures of change, performance, and project management are established and communicated.

Finally, as the organization becomes mature, the messages tell the organization how well it aimed, fired, and hit the bull's-eye. Most of the communications are very pointed, have specific content, and strive to keep the organization focused on the same bull's-eye all of the time. Recognitions emerge and comparisons to Best-in-Class organizations become commonplace.

1.8.3 Audience-Based Porpoising

This section talked about varied types of communications and the dichotomy of content in communication messages. The final consideration is the preciseness and specificity of the messages based on the types of audience. The content and delivery of a message to the general population of an organization will be greatly different from what is delivered to a Kaizen improvement team focused on a specific opportunity with a short window of improvement and solutioning. This is referred to as porpoising.

Key Messages in This Chapter

1. If you do not believe that you need to begin transforming into a Learning Organization, take one more step and do the assessment.
2. You personally must have the conviction to lead a journey of transformation for the organization. This is the most critical question. Stop now if the answer is no or not at this time.
3. Technology can be an inhibiter to Learning; it enables inefficient things to be done so fast that we ignore the time-based process waste because it is so small.
4. Simplicity is a virtue—only five steps: assess, architect/plan, prepare, do, learn.
5. Lean is basic, repeatable, and sustaining when coupled with a truly change-hungry organization.
6. Planning for a transformation does not have to be precise; there are too many unforeseen things that will cause the best plan to implode. Get the direction right and run toward it.
7. The risk of doing nothing is greater than the risk of change.
8. Winning through position is not good enough; somebody is likely going to eat your lunch.
9. Disruptive technologies and competitors are a great risk for already successful companies; if you do nothing, the risk that something disruptive will emerge is genuine and may take your market/customers away.
10. Lean is simply described as minimalist process design.
11. Learning Organization and Organizational Learning are two different things, but they are very complementary.
12. Communication is critical throughout transformation, and it must be varied in content and level based on audience type, size, and organizational maturity.

Challenge Actions

This section is about "Committing to the Need for Change." It focused on introducing the book's premise, foundation concepts on Lean and developing an organization that learns from using Lean. There is a single challenge action for this section that requires a willingness to commit your organization to embark on a journey of transformation.

Challenge Actions—Commitment	*Points*
1. Invest time to internalize the challenge and effort required to launch your organization forward into its transformation activity. Focus on what it's going to take to instilling Lean into your personal behaviors, your organization's cultural and behavioral fabric. Answer two questions; a) Are you personally committed to leading the organization toward becoming a Lean Learning Organization? b) Are you ready to start now?	N/A

Chapter 2

Change Starts with Knowing What You Have

> If you don't like change, you're going to like irrelevance even less.
>
> **General Eric Shinseki, 1942–present**
> *U.S. Army Chief of Staff (ret.)*

You are the leader of the organization and everyone looks to you, and at you, for direction, guidance, discipline, and behaviors. We are leading you through the first step to increase your understanding of what your strengths and challenges are to lead the organization through the transformation.

This chapter is all about assessing and inventorying today's "as-is" competencies, capabilities, and behaviors to know exactly what you have to begin your journey. When the inventory is complete, you will know where the most prevalent strengths and largest opportunities exist within yourself and your organization.

In Figure 2.1, the marked actions are the starting steps—recognize the need for challenging the status quo and assessing both leadership and organization behaviors and practices targeting Lean, Organizational Learning, and acting like a Learning Organization. Remember, the design of these steps is intended to be quick, simple, and relatively painless. If they get too onerous, the organization will buckle under their weight, and resistance will rise, potentially to a point where the organization rejects the action as "non-valued" and forward progress stops, even retrenches as old ways are used with greater conviction.

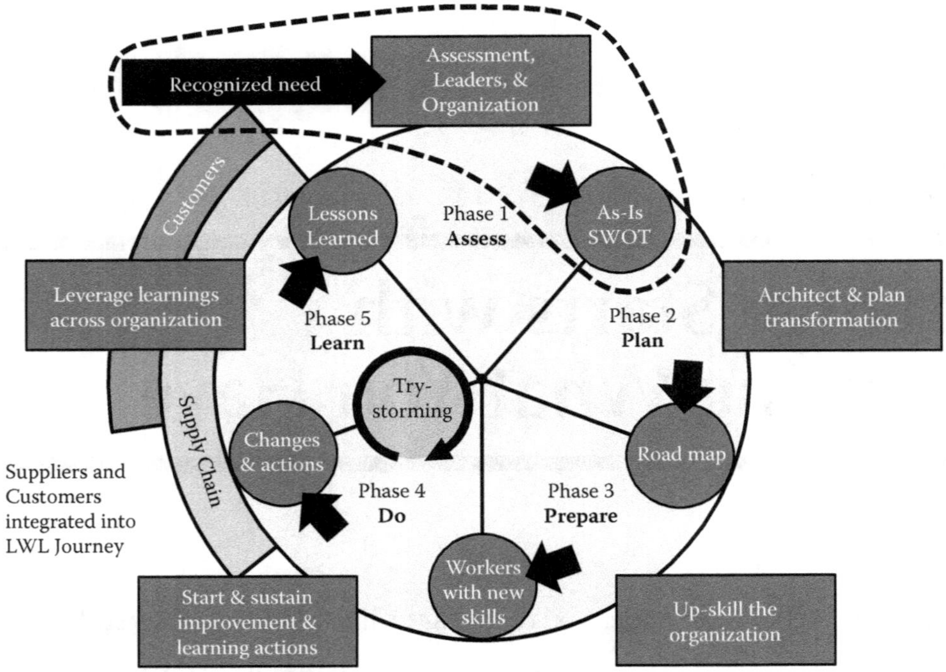

Figure 2.1 Assessment, phase 1—transformation model.

In the balance of this chapter, you will be led through various assessment instruments used to understand both the leaders' and the organization's current condition. A quick word about balance is important. You will be presented with many assessment instruments; each one can provide you with valuable information in preparation for developing your plans. It is not our intent to prescribe a regimen of assessments that will consume huge amounts of your and your organization's time. The approach strives to ensure that you get the right information, at the right speed, at the right time, and with the right amount of investment. Use judgment on exactly which assessment tools meet your needs based on the balance between investment in resources and the type and content of the output provided.

Treat problems as advantages, not disadvantages. Every time there is a problem, it allows creative people to create solutions. This achieves positive, measureable impacts to the organization and it helps bring the employees closer together. They use a single vocabulary and simple common tools, and as their maturity rises, they begin to look and solve system-level problems, not just process-level problems.

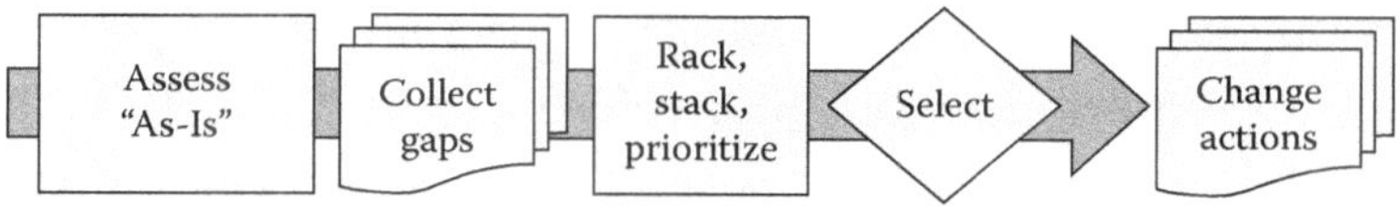

Figure 2.2 High-level rack, stack, and select gaps for action list.

2.1 Assessments Build Stronger Organizations

The assessments examine the organization's understanding and ability to use the values of Lean, and OL, as it relates to the core organization practices, the specific outputs/deliverables, and how the values of Lean mesh with the core organization values (there must be no overt or sub-textual conflicts; if there are, fix them before asking the organization).

In Figure 2.2, there is significant activity involving assessing and collecting gaps, assembly of inputs (racking, stacking, and prioritizing), and then making decisions for actions. Decisions should be made based on resource investment, precedents (some actions come first), and impact on organization results.

Later in the book, exhibits and examples of decision-making tools that are easy to use and enable fast and efficient decisions are presented. This is setting the groundwork so decisions can be empowered at lower levels in the organization, enabling a more accountable workforce and nimble organization.

The instruments contained in this book are designed to be fast in execution, provide enough detail and information at a granularity level that is immediately actionable, and not so onerous in overhead that the organization rebels. Being "fleet of foot" is the intent for their use, not exquisite, complicated, and precise execution that translates into overhead.

2.2 Learning Atrophy

> There cannot be mental atrophy in any person who continues to observe, to remember what he observes, and to seek answers for his unceasing hows and whys about things.
>
> **Alexander Graham Bell**

Learning atrophy is a person's or organization's inability to process new information from experiences or other references and use it for the betterment of the organization or oneself. For comparison, let us look at the physiological description of muscle atrophy (Sandri, 2008):

> Muscle atrophy, or disuse atrophy, is defined as a decrease in the mass of the muscle; it can be a partial or complete wasting away of muscle. When a muscle atrophies, this leads to muscle weakness, since the ability to exert force is related to mass.

In this description, muscle atrophy is defined as the decrease in muscle mass. The parallel is that an organization's learning atrophy is associated with its loss in infrastructure and culture to learn. For this book's purpose, it is defined as:

> Learning atrophy is the loss in intellectual capacity and capability (mental muscle) to rapidly capture and analyze data, synthesize it into meaningful information, and use the information to adjust/change as a result.

What is seen in organizations caught in this state are their inability to make and implement rapid changes, loss in flexibility and adaptability to changing environments, resistance to change at the emotional level, and a desire to accept the status quo.

What does it take to break free from this paradigm? Let us again look at the physiological equivalent noted by Sandri (2008):

> Muscle atrophy can be opposed by the signaling pathways which induce muscle hypertrophy, or an increase in muscle size. Therefore one way in which exercise induces an increase in muscle mass is to down regulate the pathways which have the opposite effect.

This boils down to regular and frequent management exercise that reverses atrophy and reignites the communication and management pathways (command and control practices of an organization) for increased capacity and capability to change and improve.

2.3 Take Stock of Yourself

There is no greater delight than to be conscious of sincerity on examination.

Mencius

In his book, *Out of the Crisis* (Deming, 1982), the author's Theory of Profound Knowledge points to four significant dimensions that all leaders must understand and embrace. He makes the point that true learning never takes place unless you place your credibility on the line. Alternately, true learning does not occur unless you become a real part of the learning experience. That is, to find if a concept is true, or if an idea works, you must act as though it works, follow through, and learn from both the successes and failures of that action.

Deming advocated all managers to have what he called a System of Profound Knowledge, consisting of four dimensions:

1. **Appreciation of a system:** A leader must understand the system he or she is attempting to manage. Without this understanding, the system cannot be managed or improved. A system cannot understand itself or manage itself. Optimization of the parts does not optimize the whole. System optimization requires coordination and cooperation of the parts, which requires leadership.
2. **Knowledge of variation:** This refers to Shewhart's (1931) concept of common or system causes of variation and outside assignable or special causes of variation. It pushes back on blaming people for variation caused by the system.
3. **Theory of knowledge:** Knowledge depends on understanding of theory. Information is not knowledge. Experience teaches nothing without theory. Practice makes permanent, not perfect. Copying examples does not lead to knowledge.
4. **Knowledge of psychology:** Leaders must understand human behavior to motivate, coordinate, and manage people to optimize the system.

In Table 2.1, three fundamental leadership roles emerge: designer, teacher, and steward, outlined by Senge in his book, *The Fifth Disciple: The Art and Practice of the Learning Organization* (Senge, 1990). These three roles form a backbone of how effective leaders should act and be viewed by the

Table 2.1 Leadership Roles that Fortify Learning with Lean

Role	*Characteristics*	*Fortifies Learning with Lean*
Designer	• Seen as a designer not captain • Creates a common vision with shared values and purpose • Determines policies, strategies, and structures that translate guiding ideas into organization decisions • Creates effective learning processes that enable continuous improvement of the policies, strategies, and structures	• Provides the end game, direction, and speed for change • Sets clear expectations supported with policies and strategies that align and enable individual and organizational learning to occur • Is a content expert, unique to the organization, with the ability to match learning styles and processes with organizational culture and behaviors, while targeting appropriate impacts to the organization
Teacher	• Seen as a coach • Works with mental models present in the organization • Understands the concepts of reality and restructures views to address underlying causes of the problems	• Presents him or herself as a knowledge leader and demonstrated practitioner of Lean skills • Is able to effectively translate Lean technologies across all functions, and up or down the leadership chain • Can effectively convince and enroll lay people and employees who have new or emerging skills for the purpose of continuous improvement systems
Steward	• Seen as integral to the organization • Leader feels that he or she is part of something greater • Desire is not to lead, but to serve • Focused on the greater purpose of building better organizations and the way they operate	• Seen as the lightening rod for the Lean transformation • Has the passion and the drive to fully engage with improvement teams regardless of level • Is capable of defending the approach in deployment against natural naysayers • Maintains strong presence, visibility, and communication actions that ensure successful deployment and long-term sustainability

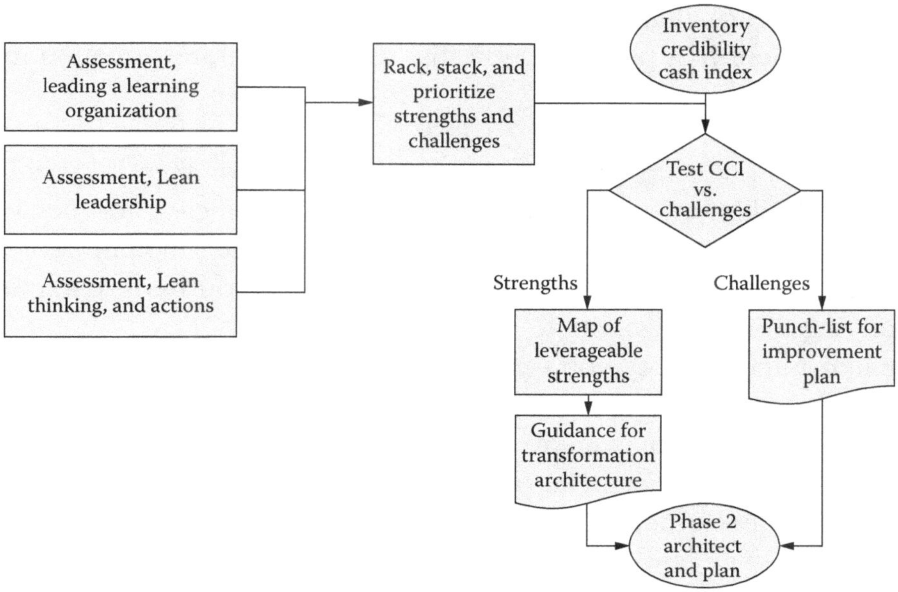

Figure 2.3 Flow of assessment actions for leadership.

organization to manage a transformation. Associated with each role are key characteristics and associated behaviors that fortify LWL.

This may seem like a lot of work to get the ball rolling, but you would never start a cross-country automobile journey without knowing the condition of your car and other resources you may or may not have to use to make the trip. This is the same condition; you need to know what your strong and weak capabilities are about leading organizational learning, Lean methods and tools, and how well equipped you are to lead a Learning Organization. In Figure 2.3, there is a high-level flowchart of assessment activities that infer precedence, timing, integration of findings, and gap identification for the organization's leadership. Two specific outputs include a personal development plan creation and a determination of how much personal change equity (Credibility Cash Index [CCI]) is going to be required to impact the inertia of the organization's current direction.

The flow combines several activities into a simple process for action:

- Multiple, separate assessments evaluating a wide range of aspects
- Integration of assessment findings into a composite "hit list"
- Rank and stack findings, application of a prioritization schema
- Selection of the top few (most critical to personal contribution to transformation success)

- Comparison of change needs with an inventory of CCI
- Development of a plan of action that will set the transformation initiative into life

The next output is an action plan that directly guides your personal activities to help build, fortify, and maintain skills, competencies, and behaviors crucial to leading an organizational transformation. The timeframe for the action plan is predicated on the direction and speed you have determined for changing the organization.

The effort and honesty you expend here will show itself many times over when you kick start the organization in the same direction. Your senior, middle, and front line management should have the same opportunity for assessment and evaluation. The opportunity to reap the benefits you are getting from your assessment can be realized for them as well.

So, how do you assess organizational learning traits, competencies, and behaviors? By using the simple assessment instruments contained in the Appendix and outlined in Figure 2.3. You can quickly gather information on strengths and challenges about yourself, the leadership team, and the organization as a whole. It is important to use caution with these instruments because they are fast and efficient tools that when used together provide coarse levels of data and information. They rely on anecdotal references, personal impressions, and qualitative summaries from a broad set of individuals that may or may not be calibrated to a common scoring methodology.

Remember, instruments like this provide a wealth of nonscientific information and should be used as just one of the series of in-feeds into the process of gap identification, prioritization, and selection. For some organizations, a few of these questions may need to have minor revisions to assure relevance to the organization. These are designed to be easily configured to best match an organization's mission, culture, behaviors, and actions.

2.3.1 Are You Committed to Leading the Change?

Up to this point, the steps, actions, and foundation concepts have been shared and reinforced. You are ready to begin the journey, but the initial step is crucial—you must be able to answer honestly these simple questions:

- Do you genuinely believe that there is a need for change?
- Are you personally ready to commit and lead the organization?

- How frozen (stuck in the present) are you?
- Are you capable of unfreezing and making personal changes?

The first question has an answer that is binary; it has no shades of grey. Either you believe that change is required or you do not. If the answer is yes, the rest of the book will make sense and will provide a recipe for guiding your change and transformation. If the answer is no, then this book becomes an academic exercise and, while filled with information and action steps, you will find that it is not immediately actionable for you and your organization.

Let us recognize that there are times when it is okay that the answer is no right now, but it will be yes in the near future. Something significant may already be happening and it cannot be disrupted. If that is the condition, read on; see the journey and actions, prepare for a launch, and know that a significant change is around the corner. However, you must decide if it is time, and if you are ready to build an organization that is change-hungry and possessing the capability for a sustainable competitive advantage.

From this point forward, the book is constructed based on you providing yes answers to the initial questions about your personal belief that change is needed. The implications of these answers are:

- You believe that change is required.
- You are committed to starting a journey.
- You are willing to be open, honest, and precise in assessing the current state.
- You will be the lightning rod for all of the effects to the organization.
- You will critically evaluate what is required to change and invest in skills, tools, and methods accordingly.
- You will lead by example the behaviors, actions, and culture you expect in others.

These are not negotiable points. Your commitment to the transformation for your organization, your knowledge of the direction, broad course of actions, and speed at which you would like to change should be well understood.

Never debate about the end game—always debate about the rate of change! That is what is crucial to the transformation. Challenge yourself, your leadership team, and every person in the organization to use critical examination for tattletale signs of the rate of change and how it is taking root. Table 2.2 provides a few simple examples of tattletales (visible evidence) that are typical of organizational change. One of your challenge

Table 2.2 Tattletale Examples of Positive Organizational Change

Simple Tattletales	*Visible Evidence*
Spontaneous waste elimination activities	Without leaders or managers directing process streamlining, waste identification and removal activities are present across the organization.
One-Touch Flow (OTF) evident	Every employee (top to bottom) uses the concepts of OTF as they analyze their work, and their language reflects the concepts and communications.
Visible measurements	Measurement of process performance that is important to the organization is accomplished and visible (physically or virtually) to everyone involved.
Using variation versus conformance to measure performance	A clear shift in how processes are being measured now focuses on examination of process output variation, comparisons to true requirements, and determination of margins of performance are used on critical processes.
Management Standard Work (MSW) evident	MSW actions are in place across the organization including performance review and analysis, closed-loop corrective action effectiveness, and leveraging learnings across organizational boundaries.

actions at the end of this chapter is to build a unique set of tattletales for your organization. Use specific examples and language that everyone can use, sort of a cheat sheet to know when a change is happening.

Being a leader and being committed means you have to be on the work floor, seeing how work is actually being done in order to effectively understand, learn, and lead. This is not the concept of Management by Wandering [Walking] Around (MBWA) (Kouzes & Posner, 1988), but it is your demonstration of behaviors about involvement in work actions, learning and knowledge sharing of focus on identification and removal of waste, and daily recognition of involved workers and their contribution to the organization's successful mission. An unattributed quote that is commonly referenced sets the framework well regarding managers', senior managers', and leaders' knowledge of how work actually is done.

> "… the higher you are, the farther away from the process, the less you know …"

Table 2.3 Attributes of Level 5 Effective Leaders

Professional Will	*Personal Humility*
• Creates superb results • Clear catalyst in the improvement transition	• Demonstrates a compelling modesty • Shuns public adulation, never boastful
• Demonstrates an unwavering resolve to produce the best long-term results in spite of all difficulties	• Acts with quiet, calm determination • Relies on inspired standards, not inspiring charisma, to motivate
• Sets the standard of building an enduring great organization • Settles for nothing less	• Channels ambition into the organization • Sets up successors for even greater success in the next generation
• Looks in the mirror • Apportions responsibility for poor results • Never blames other people, external factors, or bad luck	• Looks out the window, not in the mirror • Apportions credit for the success of the organization to other people, external factors, and good luck

The evidence for the validity is right outside your office. Pick a common process; go to where it is executed and "walk the flow." What you see will assuredly be very different from what you know and how you think it is executed. The disconnect most often is between how it was designed and approved by management for implementation and how it is actually executed on the floor.

In the book *Good to Great* (Collins, 2001), the author supplies four groups of attributes for Level 5 Executives using professional will and personal humility. Table 2.3 brings home behaviors that are essential for leaders who want to successfully lead organizations through transitions into being great organizations.

One of the first assessments needed of the leader is one that evaluates the strengths and the challenges that you have in leading toward implementation of OL. OL, as defined in Chapter 1 and for the purposes of this section, talks about the skills, behaviors, and attitudes to be able to establish and reinforce good learning behaviors within the organization.

It will seem like the many assessments you are doing as a leader cover much of the same ground; in fact, each one is evaluating various skills, capacities, capabilities, and cultural norms that you show to the organization as the leader in your ability to effectively initiate and lead the transformation.

The rate of change is the critical factor. Without a concerted drive by the leader to understand the changes in the organization, work, and deliverables, the organization will not see its leadership team with personal investment ("skin in the game")—there is no personal cash in the bank that they can use to direct and enable change.

Using Lean as a vehicle to intimately see and know the organization is one of the fastest ways to build up quickly the leader's bank account (*see* CCI, Section 2.2.5). Lean's basic premise is to get on the floor, walk the flow, see the trials and issues, and learn what does and does not work.

Another facet of being an effective leader is getting out of the trap that errors, or waste, mostly come from other parts of the organization. A true leader is one that first looks internally and then externally. In the beginning, examine your core processes, uncover the errors and waste, determine their root causes, and eliminate the contributors. Then start the next layer of streamlining by looking at the processes' primary in-feeds. Use what you discovered in streamlining your processes as key learnings and develop a synergy with other input systems and processes to accelerate their rate of improvement and change. Ultimately, as this "learn–observe–partner–teach" deployment approach becomes more commonplace across the organization, it builds credibility and effectiveness for sustaining an accelerated rate of change in the organization. This borders on the brink of becoming a cultural shift and enabling sustainable change behaviors—both key to moving toward demonstrating a sustainable competitive advantage.

2.3.2 Check Your Ability to Lead a Learning Organization

The first assessment is about evaluating the competency practices that you demonstrate in Leading a Learning Organization. Its focus is to evaluate and assess how well you promote, recognize, and leverage the 10 characteristics of leadership roles that fortify a Learning Organization.

Exhibit 2.1 shows only the first page of an instrument that uses subjective scoring with a non-calibrated group of assessors. This group should include your manager, peers, partners, and employees. The details of its scoring methodology, its calibration, and its interpretation are contained in the Appendix, Exhibit A.1.

Assessment – Leading a Learning Organization

Scoring:
1 = Little/ no evidence of leader behaviors/actions
2 = Some/ ad-hoc evidence of leader behaviors/actions
3 = Occasional/ often evidence of leader behaviors/actions
4 = Most of the time there is evidence of leader behaviors/action
5 = Consistent/always leader behaviors/actions

	Leader Attributes – Learning Organization	Scores 1	2	3	4	5	Scores by Role
Designer	Provide the 'end,' direction and speed for change				×		4.0
	Sets clear expectations supported with policies and strategies that alien and enable individual and organizational learning to occur					×	
	Is a content expert, unique to the organization, with the ability to match learning styles and processes with organizational culture and behaviors, while targeting appropriate impacts to the organization			×			
Teacher	Presents themselves as a knowledge leader and demonstrated practitioner of Lean skills			×			3.7
	Is able to effectively translate Lean technologies across all functions, and up or down the leadership chain			×			
	Can effectively convince and enroll lay people employees, who have new or emerging skills, for the purpose of continuous improvement systems					×	
Steward	Seen as the "lightening rod" for the Lean transformation			×			3.3
	Has the passion and the drive to fully engage with improvement teams, regardless of level				×		
	Is capable of defending the approach in deployment against natural naysayers			×			
	Maintains strong presence, visibility, and communication actions that ensure successful deployment and long term sustainability			×			

Exhibit 2.1 Assessment—leading a Learning Organization (example).

2.3.3 Check Your Lean Leadership Capabilities

Let us do a quick check. Thus far, you have assessed your abilities to lead OL and lead a Learning Organization. The third leg in the assessment triad is one that evaluates your Lean leadership capabilities. This brings to the table a need to understand your knowledge of Lean tools and practices and

Table 2.4 Dimensions of Lean Leaders

	Project Implementation	*Ongoing Operations*
Passion for Lean	Passionate about the potential for Lean to make the enterprise successful and fulfilling.	*Same plus* willing to make personal changes in one's own work, including using personal standardized work.
Disciplined adherence to process	Sets expectations, regularly uses a process to track and follow up on actual accomplishment of tasks.	*Same plus* exhibits intense commitment to focus on explicitly defining and executing processes.
Project management orientation	Prior experience in successfully implemented projects; uses a defined process to track performance, completion of tasks, and corrective actions.	Able to identify needed changes based on timely process data; assigns tasks leading to successful changes; uses visual processes to track, follow up, and take appropriate corrective action.
Lean thinking	Understands Lean concepts; has experience applying Lean concepts; actively promotes a Lean future state; applies Lean concepts in daily project work processes.	Actively uses goals to set improvements; uses One-Touch Flow consistently; coaches root-cause orientation to corrective action; knows process improvement methods; is able to personally lead Lean improvements.
Ownership	Thinks and talks about the area as his or hers to lead, set direction for, change, and improve.	*Same plus* eager to empower others through structured implementation of their ideas; acknowledges and celebrates improvements made by all.
Tension between applied and technical	Understands the need to sweat the details, as well as to get things done; is willing to listen to technical experts and consider their advice in planning for the implementation.	Understands and respects the details; actively supports upgrades to performance and exposes previously hidden impediments; uses a "what can we do today" change orientation.

Table 2.4 Dimensions of Lean Leaders (continued)

	Project Implementation	*Ongoing Operations*
Balanced commitment to production and management systems	History of effective give and take with people and ideas; evidence of process focus beyond a "hit the numbers" approach to management; eager for greater participation by affected people.	Personally treats process focus as crucial to success; can "see" waste and opportunity even in Leaner processes; insists on visually tracking performance and execution; insists on analysis and appropriate, timely corrective actions.
Effective relations with support groups	History of getting things done with support from operations support groups such as engineering, quality, production control, safety, finance, HR.	Understands roles, responsibilities, and expertise of support groups in plans for improvement and responses to problems; makes expectations explicit for support group performance.

asks questions of how you demonstrate this knowledge and behaviors in project and operational level work. In his book, *Creating a Lean Culture* (Mann, 2005), the author describes Lean leadership characteristics using eight dimensions applied to two types of accountabilities.

In Table 2.4, eight dimensions for Lean leaders are identified, starting with passion for Lean and ending with effective relations with support groups. These dimensions are also recognized for their differences in two functions: project implementation and ongoing operations.

In ongoing operations, there are several cells that are noted as "*same plus*," implying that all of the attributes in the project implementation column are relevant plus the new additions noted. These eight dimensions are used to evaluate Lean leadership.

In Exhibit 2.2, an example of the simple assessment instrument can be seen. It is broken into two topical streams aligned to Table 2.4, for project implementation and ongoing operations. There are eight areas of Lean leadership behaviors and each one can be graded from 0 to 5. The grades are subsequently combined and the results reflected in the two polar (spider) charts at the top of the instrument. Through examination of the behaviors with high scores, especially those with strong relationships to each other (e.g., disciplined adherence to process versus balanced

Check - Lean Leadership Self Assessment

Purpose: Assess Leadership's Lean Culture and Competency
Scoring: 0 N/A - cannot score
1 Little/no evidence, avoidance of Lean
2 Some/ad-hoc evidence, slow adoption
3 Some systematic evidence, no learning
4 Lean systems evident, few cycles of learning
5 Lean systems and behaviors are key drivers

Color Code for Spider Charts (below)

- Passion for Lean
- Disciplined adherence to process accountability
- Project management orientation
- Lean thinking
- Ownership
- Tension between applied and technical
- Balanced commitment to production and management systems
- Effective relations with support group

Lean Leadership for Project Implementation

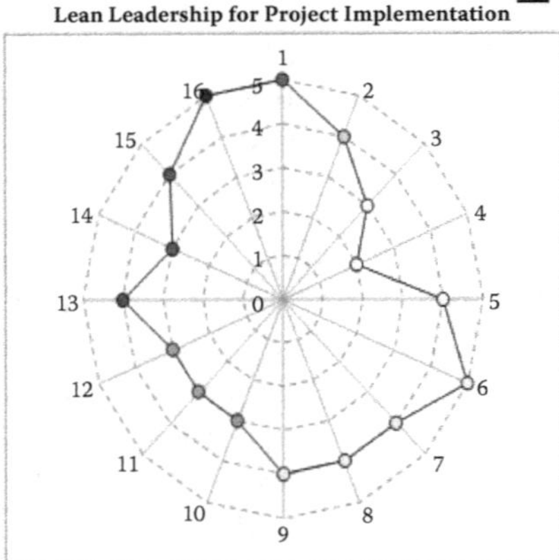

Lean Leadership for Ongoing Operations

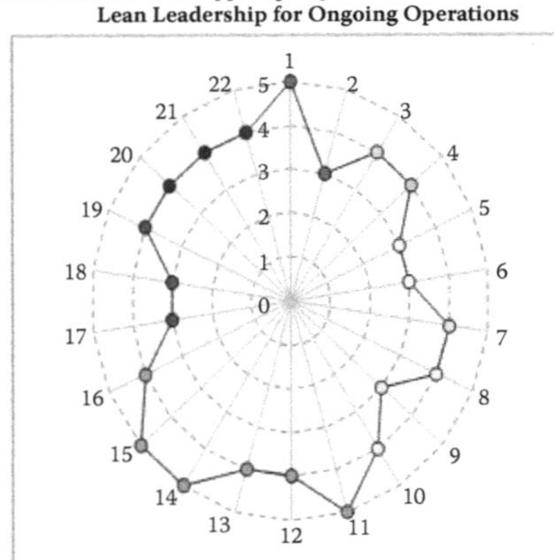

Attribute	For Project Implementation	0	1	2	3	4	5	For Ongoing Operations	0	1	2	3	4	5
Passion for Lean	1. Passionate about the potential for Lean to make the enterprise more successful and work more fulfilling for all involved.						×	1. Passionate about the potential for Lean to make the enterprise more successful and work more fulfilling for all involved.						×
								2. Willing to make personal changes in one's own work, including using standardized work for his/her own position.				×		
Disciplined adherence to process accountability	1. Sets expectations, regular uses a process to track and follow up on actual accomplishment of assigned tasks.					×		1. Sets expectations, regular uses a process to track and follow up on actual accomplishment of assigned tasks.					×	
								2. Exhibits intense commitment to focus on explicitly defining processes and disciplined adherence to them.					×	
Project management orientation	1. Prior experience in successfully implemented projects.				×			1. Able to identify needed changes based on daily process data and assign small-bite daily tasks leading to successful implementation of the changes.				×		
	2. Uses a defined process to track performance and completion of task assignments.			×				2. Uses explicity defined visual processes to track and follow-up assignments and take appropriate corrective action.				×		
	3. Identifies corrective action where necessary and follows up on it.					×								

Exhibit 2.2 Assessment—Lean leadership (example).

commitment to production and management systems), patterns of strengths and opportunities emerge quickly. These become in-feeds to development of personal improvement planning.

Exhibit 2.2 and Exhibit 2.3 show only the first page of an instrument that uses subjective scoring, with a non-calibrated group of assessors. This group

Check - Lean Thinking & Action Self-Assessment

Note: Place an X in only ONE score box for each question (light yellow cells below).

	Question	Strongly Disagree 1	Scores (*Enter score in ONLY one*) Disagree 2	Neutral 3	Agree 4	Strongly Agree 5
1	When we go into a decision meeting or project review, we know that the right people will be in the room to make decisions.					
2	Managers in my organization ask good, challenging questions when I present my findings and recommendations to them.					
3	When I recognize a problem that we have seen before, I have to ability to find out how we solved it last time.					
4	We explore multiple alternatives before making key decisions.					
5	Subject matter experts (SMEs) get regular opportunities to deepen their customer knowledge with early customer feedback on their process designs.					
6	Process execution and reviews meet their schedules consistently.					
7	Decision makers take the time to understand the problems, alternatives and recommendations before making a decision.					
8	There are few loopbacks late in an improvement action's implementation cycle.					
9	We take the time to capture what we've learned so that we can share it with others and reuse it ourselves later.					
10	We do not hold up the flow of improvement work with artificial barriers imposed by reviews or decision meeting.					
11	We take some time to understand root causes before we recommend countermeasures or solutions.					
12	Customers are highly satisfied with our improvement action's quality.					
13	When we go into a decision meeting or review, we are confident that there will be no last minute objections or issues to delay decisions.					
14	Managers in my organization use systematic problem-sloving to solve problems within their span of control.					
15	We take the time to identify the people who need to be consulted about a decision, and then engage them in the decision-making process.					
16	We take the time to measure results and reflect upon the effectiveness of the decisions that we make so that we can learn.					

Exhibit 2.3 Assessment—Lean Thinking and Action (example).

should include your manager, peers, partners, and employees. The details of its scoring methodology, its calibration, and its interpretation are contained in the Appendix, Exhibit A.2.

2.3.4 Check Your Lean Thinking and Actions

You have covered your leadership of OL and leading a Learning Organization. Now we are going to evaluate Lean Thinking and Actions as a core set of competencies.

This is a reminder that while taking several instruments to evaluate different competencies for leadership, these tools are only guidance instruments. Their intent is to get you thinking about what should be routine actions across the leadership team and the organization.

In this case, the instrument focused on 24 questions. Each question is independently evaluated and scored using subjective scoring and can be assessed at individual, peer, and manager levels.

Thus far, you have completed assessments on leading OL, leading a Learning Organization, Lean leadership, and Lean Thinking and Action. You would think this would be enough data and information to make effective personal improvement plans, but there is something missing. It is the evaluation of your personal credibility within the organization to make an impact, through control, influence, or history. In the next section, you will be exposed to the evaluation process that inventories the quantity of leadership influence called CCI.

2.3.5 Check Your Credibility Cash Index

When you arrive in an organization, your credibility "account balance" is just about zero. The organization has no experience with you; there is no track record of success built on employee–management partnership. You do not bring a complete knowledge of how the organization's work is done to achieve the deliverables and services experienced by its customers. The organization genuinely does not know what you have done in the past or how you did it. You may have some "positional cash" (you are the senior leader of the organization) and that is about it.

Think of the CCI as an "influence" checking account—it has a balance and requires deposits to build the balance to the level where you can use it (write "improvement action" checks to withdraw the cash) for enabling process changes and improvement.

For clarity, the CCI is defined as:

> An integrated measure that gauges a person's inventory of organizational influence and control that enables effective leadership for desired organizational changes.

Five components make up the CCI's method to gauge the balance in your account. The index is a simple integration of scores for the components: Change History (CH), Results History (RH), Organizational Ownership (OO), Organizational Influence (OI), and Conviction for Change (CC). Table 2.5 briefly describes each of these elements and what makes them up.

You must first start adding to your account. These additions to your account come from rapid-fire, successful improvements that make a positive impact to the organization. These must be mixed with changes that improve the localized activities making employees' work tasks easier and more productive. Employees will quickly see pockets of excellence in target organizations grow into full deployment examples. Your leadership passion and commitment for change must be easily heard and understood. By rolling up your sleeves and making change happen, the account balance grows to the point where you can now start investing in larger-scale change initiatives.

Table 2.5 Components of Credibility Cash Index

Change History (CH)	Demonstrated evidence that you were instrumental in the design, implementation, and leadership of organizational change toward being a Learning Organization.
Results History (RH)	Visible positive changes to organization-critical measures that are linked to the changes in organizational design, methods, and discipline.
Organizational Ownership (OO)	Leader has defined ownership, accountability, and authority to make and lead the changes being implemented.
Organizational Influence (OI)	Leader has established influence with key owners and stakeholders responsible for parts of (or the whole) organization where changes are being implemented.
Conviction for Change (CC)	There is clear focus, passion, dedication, and confidence of the leader to start, execute, and complete the transformation roadmap for a Learning Organization.

Employee: ____________________

	Score	Scoring Criteria 1	2	3	4	5
Change History (CH)		No evidence of that their organization changed through Lean learning	Some	Many change efforts show successful transformation and use of Lean learning	Most	Every organization they managed learns and changes though systemic Lean improvement
Results History (RH)		No evidence of positive results from changes	Some	Clear evidence that many changes have positive results	Many	Changes always produce positive, sustainable results
Organizational Ownership (OO)		No organizational ownership	N/A–Cannot score partial ownership			Owns the organization they are trying to change
Organizational Influence (OI)		Cannot influence organizational change	Few	Many examples of influence in leading organizational change exist	Many	Clear, consistent influence of other organizations though the use of Lean
Conviction for Changes (CC)		Does not show any conviction/passion for creating change	Slightly	Generally positive and passionate towards organizational change	Almost always	Always shows, speaks and shares their conviction to learn and change

Total: ________

Exhibit 2.4 Inventory, Credibility Cash Index (example).

As expected, the higher the balance, the greater your ability to invest in organizational change. Small changes take almost nothing from the account; larger improvement actions require greater investment. You can spend up to a certain level before you need to replenish it. The cycle of spend/replenish becomes an operating norm for the organizational leadership team.

In Exhibit 2.4, you can see the five simple components that require grading using the scores defined in the Appendix, Exhibit A.4. While the scores are subjective, they can quickly and easily be aggregated from your peers, partners, or subordinates. While the scores provide a qualitative evaluation of your credibility for change, the design of the instrument is to be a fast approach to determining your level of influence and change capability within the organization. You can compare your scores with those of your peers to find your strengths and gaps in being an effective change leader.

Because the tool used to measure CCI is subjective, the values obtained are not absolute measures, but are a relative gauge of how someone is seen by others. It provides an indicator of how much influence he or she can have on an organization to make a change. Figure 2.4 shows the relationship

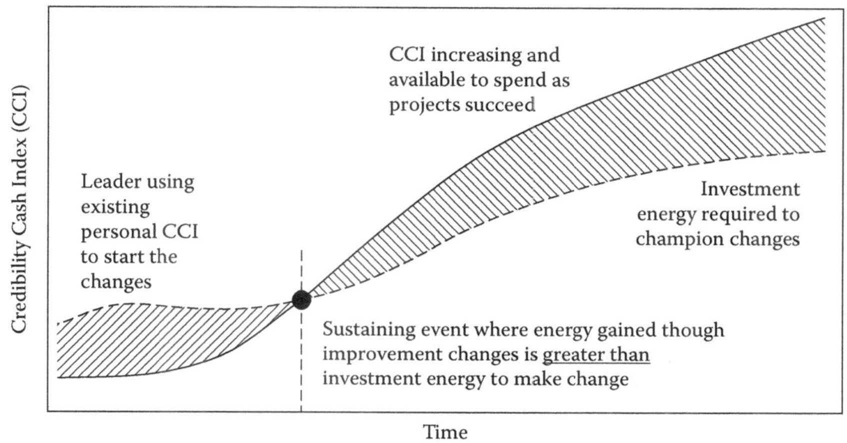

Figure 2.4 Credibility cash index requirements change with time.

between the ability to effect changes when someone does not have the personal clout or assets in the beginning of a journey and when they have acquired a higher level of personal clout by earning the organization's value of them.

Typical questions that could be going through your mind at this point include:

- How much personal "cash" do you have?
- How much can you afford to spend on the transformation in light of what you need to run the organization?
- How much is required to change from today's current state to a future state where a Change-Hungry Organization is assured a sustainable competitive advantage?
- What do you need to do to add or borrow personal "cash"' from your account so you can start changing the organization?

Each of these questions is probative. Simply by taking the CCI inventory, you will answer the first. The second question is dependent upon how many changes you are trying to do simultaneously. The answer to the third question lies in between the current level of inventory, the desired end game, and the results of the assessments to tell you the size of the gap or what you have to bridge. The last question is best answered by example—in the beginning, you will need to use some of your credibility cash to get started. Additionally, you will be adding to your account every time there is a win or success. This is a natural by-product of making changes that impact the organization or the work life of the employees.

Every project uses up CCI. How much can you afford to spend? Know that as projects are done, cash builds up in the bank again so you can spend it on other projects and initiatives.

The leader is the beacon for the organization—his or her behaviors become the reflection of the organization. The leader must want to learn and must want to transform learnings into changes. Leaders in organizations that use Lean as a vehicle exhibit several specific behaviors in how they drive their organization (Mann, 2005).

With the understanding of how much Credibility Cash a leader has in his or her bank account, it is time to assess and determine the strengths and opportunities or challenges presented within the organization. There are equivalent tools that the organization will use to evaluate its as-is state.

While the outcomes from the assessments will have different contexts, they will align and point out significant gaps between leadership and the workforce. There are differences in expectations, the natural byproducts of organizational growth and maturity, where the leaders and the workers see dramatically different landscapes. The purpose of Phase 1 of the transformation model is to identify and vet the differences, and then use them as opportunities to develop the transformation change plan.

2.4 Take Stock of the Organization

> Once we know our weaknesses they cease to do us any harm.
>
> **George C. Lichtenberg**

Much like the evaluation you performed on yourself and your leadership team to understand what their strengths and opportunities are, the organization needs to be evaluated for its behaviors and practices regarding the way it learns, how fast it learns, and how it uses Lean practices to increase performance effectiveness and efficiency. In the following sections of this chapter, you will be led through assessments of the organization for Lean Thinking and Actions and an Organizational Performance Quotient.

2.4.1 Looking for the Capability to Learn

The purpose of using assessments is to evaluate an organization's capability and readiness to learn and change. They help determine the types and magnitudes of the gaps, challenges, and opportunities. The understanding enables design of the length and composition of the roadmap and milestones. Additionally, this helps identify the skills, quantity of resources required, and the likely impact to organizational performance as deployment rolls out.

You can see the similarities between the flow of actions for the organization (Figure 2.5) and those for the leaders (Figure 2.3). They both use assessment instruments, one the same, the others different, to find gaps and then have a series of steps to prioritize, select, and begin implementation of solutions. This flow is just to get the ball rolling; re-craft as needed for your organization's specific requirements.

2.4.2 Organizational Beliefs, Knowledge, and Ability

Segmentation of capabilities at leadership team, middle/front-line management, and individual contributor levels is instrumental in knowing where the organization is strongest and most vulnerable with its employees. Additionally, capabilities should be segmented by function. In doing so, you

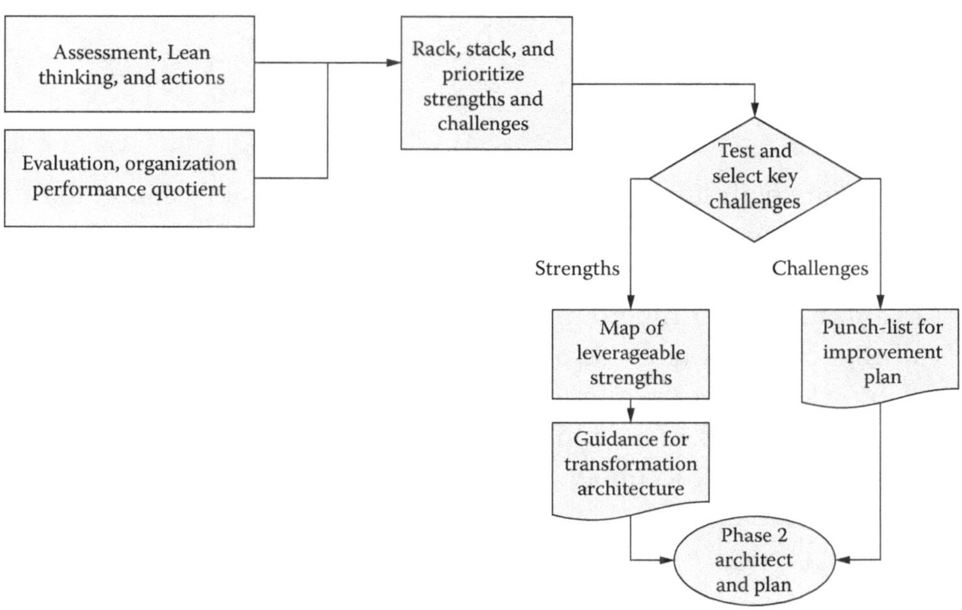

Figure 2.5 Flow of assessment actions for the organization.

have the ability to cross-compare and determine where and at what management level opportunities exist to accelerate improvement in becoming a Lean-based Learning Organization.

Three foundational attributes demonstrate an individual's or organization's capability to learn. Think of them as motive, methods, and means.

- Belief that learning is vital to each individual and the organization as a whole (motive),
- Knowledge of how to observe, challenge, and fairly determine what needs to be changed for improved performance (methods), and
- Ability to use effectively roles and responsibilities (R&R) and authorities and accountabilities (A&A) to make change happen (means).

2.4.3 Characterizing Learning Organization Dimensions

A crosscutting assessment instrument evaluates the five dimensions of a Learning Organization and enables thematic comparisons with the output from the OL assessment. The five dimensions are:

- systems thinking
- personal mastery
- mental models
- building shared vision
- team learning

The fundamental difference between the dimensions of OL and the dimensions of a Learning Organization is:

> Organizational learning dimensions focus on six areas of tactically specific actions to be able to capture, integrate, and transfer knowledge between individuals, teams or work units, and the whole organization, while the five dimensions of a Learning Organization seek to understand and evaluate cultural habits or practices that are rooted in an organization's overall culture and behavior models.

These dimensions for a learning organization are identified and discussed in Table 1.2 in Chapter 1.

2.4.4 Check for Lean Thinking and Action Competencies

This next section is designed to be able to determine the strengths and capabilities in the organization for demonstrating Lean behaviors. This is the same assessment used by the senior leaders of the organization. This time, the entire organization uses it, integrates the scores, and segments the data as needed to characterize strengths, challenges, and opportunities. Refer to Section 2.2.5 for an exhibit and brief discussion.

2.4.5 Check the Organization's Performance Quotient

Introducing the analogy between the organization and the human body allows the use of metaphors throughout the book. This starts with recognizing the basic needs of an organization and builds into the higher levels of needs. In addition, this enables the use of performance measurement, management, and the equivalent of biofeedback examined in the book, *Your Vitality Quotient* (Earle, Imrie, & Archbold, 1989).

The authors describe three factors: performance, health and well-being, and management. Each of these factors has the ability to provide biofeedback to the individual so that he or she knows how to better optimize personal performance.

Analogous to this is the concept that similar organizational factors, when applied to an organization, can also provide feedback on performance and inputs on how to optimize performance.

Roughly speaking, drawing analogies for the factors leads to these parallels:

- performance is equivalent to output,
- health and well-being is equivalent to resilience and adaptability, and
- management is equivalent to control and improvement.

This leads to a synthesis of an equivalent feedback system called EECO-feedback. EECO represents:

- Effective—Hit the target requirements of the deliverables.
- Efficient—Hit the target cost for the design and execution of the deliverables.

- Consistent—Hit the target variability allowed in the repetitive delivery of products/services.
- Organization—The target organization (team, unit, division, or whole).

In Table 2.6, these parallels are refined with definitional descriptions.

This is an important concept in OL and the drive to transform into becoming a Learning Organization. Feedback mechanisms are the single biggest link into continuous improvement. Without feedback, it is almost impossible to know and understand performance levels and improvements. It is often described as closed-loop corrective action—this is where it lands.

So, how does this translate into something meaningful to you as senior leader of the organization? Let us use Table 2.7 to explore the parallels between biophysical actions and organizational actions.

In the table, five simple physiological vitality traits are explored that provide measureable insight into how an individual is managing the three factors. These traits are:

- resistance to flow
- muscle strength
- speed of response
- recovery to normal state
- agility

These five traits are used to measure a person's physical vitality. Parallels can be drawn between a person's physical vitality and an organization's performance vitality. In the first two columns of the table, you see physiological measures and their implications for an individual, and the correlation between the measures and the vitality implications. Drawing parallels in columns three and four, you see organizational measures of performance and correlated organizational implications. Organizational measures are analogous to the physiologic measures but tailored to represent a critical deliverable of an organization. Those measures, when viewing the implications column, answer the "So what?" question for the organization.

Organizationally, when combining these, an index of performance vitality becomes available as a performance gauge. This is a useable measure, indicative of the health of the organization and its ability to create and sustain a competitive advantage.

Table 2.6 Comparison of Biofeedback and EECO Feedback

Factors	*Biofeedback*	*EECO Feedback*
Performance = output	Work delivery	Product/service delivery
Health and well-being = resilience and adaptability	Ability to withstand prolonged physical stress and the ability to adapt to the environment	Ability to withstand prolonged organizational stress and the ability to change the structure and processes to adapt to the environment
Management = control and improvement	Uses physical triggers (feedback) to enable adjustment in physical stress in delivery of work	Uses operational triggers (feedback) to drive continuous improvement, removal of wasted actions, and challenge the status quo

The vitality quotient for an individual is described in Equation (2.1), and provides a relative index of a person's vitality when compared to his or her actual age. The target range for this equation is when the vitality quotient is greater than 1.

Mathematically, the Personal Vitality Quotient (PVQ) is defined as:

$$PVQ = \frac{\textit{Actual age}}{\textit{Average of calculated ages}\,(RTF + MS + SOR + RTN + AG)} \quad (2.1)$$

Graphically, this quotient resulting from a calculated age and an actual age can be seen in its target zone using Figure 2.6. There are two examples, one in each zone above and below the "Line of Quotient Unity." The ideal quotient is greater than unity (above the unity line, in the upper left zone). Simply, the example in the desired zone represents a condition where a person's actual age is greater than his or her calculated body age—this is goodness (i.e., their physical vitality is younger than their actual age).

There is an important aspect of using this approach to gauge either personal or organizational vitality. Contained in their book, *Your Vitality Quotient*, the authors make important index translations for each of the five traits using an indexed body age based on a particular value of a specific trait. For example, a systolic blood pressure reading (Resistance to Flow trait) of 124 is indexed to a body age of 27 for men and 41 four women.

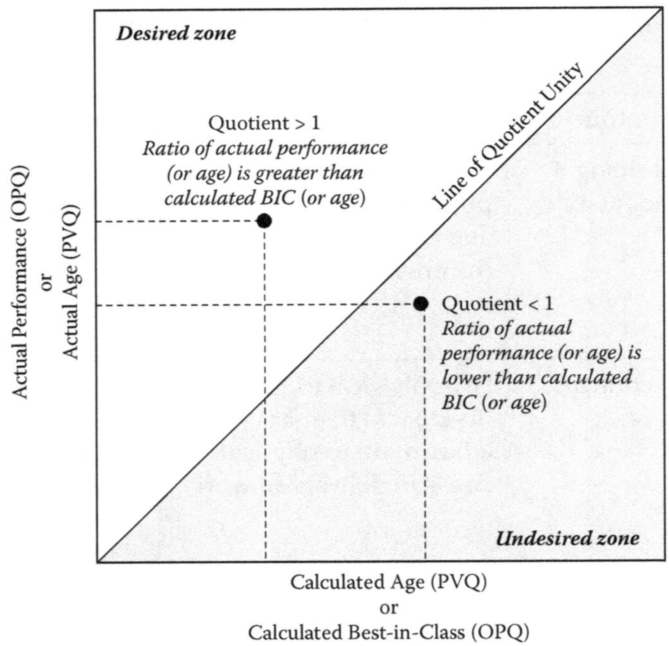

Figure 2.6 Personal vitality quotient parallels operational performance quotient.

This indexed body age is used as part of the calculated values for a PVQ in Equation (2.1).

Let us extend the approach using similar logic for an organization's performance and vitality, referred to as the Organizational Performance Quotient (OPQ). The OPQ is reconstructed mathematically in Equation (2.2) using five system traits of organizational vitality noted in Table 2.7.

This same approach must be implemented when using the OPQ. Each unique indicator used for each of the five traits, which are particular to every organization, must be indexed (similar to the methods referenced in the book, *Your Vitality Quotient*) to provide values that can be averaged in Equation (2.2). The unique challenge for this equation is the fact that there is no true chronologic age to be used for comparison. Literal replacements for chronological age, when it comes to operational performance, are Best-in-Class (BIC) measures and industry standards. These are seen in Equation (2.2) with the subscript "B" next to each of the five traits (i.e., RTF_B represents Best-in-Class Resistance to Flow measure). Additionally, each measure of operational performance when compared with its representative optimum value should be weighted according to relevance and importance to customer experience.

Table 2.7 Physiological versus Organizational Vitality Measures and Implications

		Physiological Measure	*Physiological Implications*	*Organizational Measure*	*Organizational Implications*
System Vitality Traits	Resistance to Flow (RTF)	Systolic blood pressure (BP number)	Arterial blockage to blood flow	On-time-delivery	Customer satisfaction
	Muscle Strength (MS)	Resting heart rate (HR count)	Heart muscle strength to flow more as needed	Delivery/costs	Delivery throughput margins
	Speed of Response (SOR)	Reaction time (time to react)	Neuro-musculature system health	TPT for NPI	NPI ramp, micro-responsiveness to micro changes
	Recovery to Normal (RTN)	Skin elasticity (time to recover)	Tissue recovery to normal state	Responsiveness to unplanned demand	Minimizes over-production, minimizes excess resources
	Agility (AG)	Near-vision blurring (distance)	Tissue flexibility	Deliverables flexibility (adds/drops) matching market changes	Portfolio flexibility/ response to macro changes

A side note: BIC values will not represent 1.0 in the indices, which is reserved for benchmarks from world-class processes, unless the BIC values are one and the same with world-class values.

$$OPQ = \frac{\text{Average actual performance}\,(RTF + MS + SOR + RTN + AG)}{\text{Average calculated BIC performance}\,(RTF_B + MS_B + SOR_B + RTN_B + AG_B)} \tag{2.2}$$

How do you use this OPQ equation to your advantage? Almost by definition, successful organizations most likely have reasonable performance measures in place. However,

- Are they measuring the right stuff?
- Do the measures cause change actions?
- Are the measures responsive to changing environmental needs?
- Would the customer care if the measure rises or falls?
- Are the measures used to make organization-critical decisions or change the strategies of the organization?

All of these questions represent only a small sample of what could be asked. In Table 2.8, you see measures and attributes for organization performance identified in columns one and two.

To make these timelier, impactful, and effective, the organizational measures must be translated into real-time metrics. In column three, you can see representative examples of real-time measures used to gauge performance. Additionally, column four notes where these measures would most likely be accomplished. Some reside within the organization, while some may reside in their customers' operation.

What is the value of real-time measures? It is all about the speed of information flow from source to the decision-maker. The faster that information flows, the more timely the decision, and ultimately the more impactful the decision is to both organization performance and the customer experience.

Using identified real-time measures from Table 2.8 you can construct a very simple spreadsheet that integrates these measures to create your OPQ. An example of an OPQ is seen in Exhibit 2.5. It contains sample measures for system vitality traits of a product production-based organization.

In the OPQ example, you see measures identified with "goodness" direction specified, actual values and their indexed values, plus BIC values and

Table2.8 Organization Vitality Metrics versus Real Time Metrics

		Organization Measure	*Organization Attribute*	*Real-Time Measure*	*Where Measured*
System Vitality Traits	Resistance to Flow (RTF)	On-time delivery	Customer satisfaction	On-time delivery	Customer deliverables from "core" processes
	Muscle Strength (MS)	Delivery/costs	Delivery throughput margins	Delivery/costs	Supply-chain and logistic processes and systems
	Speed of Response (SOR)	TPT for NPI	NPI ramp, micro-responsiveness to micro changes	NPI OTD	NPI development, prep and launch projects
	Recovery to Normal State (RTN)	Recovery from unplanned demand	Minimizes over-production, minimizes excess resources	Level loading, capacity utilization	Delivery, planning, and support core processes
	Agility (AG)	Deliverables flexibility (adds/drops) matching market changes	Portfolio flexibility/ response to macro changes	Time to absorb big shifts in organizational deliverables	Organization's deliverables portfolio

Organizational Performance Quotient

Function: ____________________

System Vitality Trait		Measure	Actual Value	Indexed Value	Best-in-Class (BIC)	Indexed BIC
	RTF (resistance to flow)	OTD (on-time delivery [up = good]	92%	0.93	99.50%	0.99
	MS (muscle strength)	Cost/delivery [down = good]	$12K/ 10,000 lbs.	0.82	$11.3K 10,000 lbs.	0.95
	SOR (speed of response)	NPI OTD [up = good]	89%	0.91	96%	0.98
	RTN (recovery to normal)	Capacity utilization [+ target = good]	94%	0.96	92%	0.99
	AG (agility)	TPT (through-put time) for large change [down = good]	8.1 months	0.72	6.6 months	0.91
			Averages:	0.87		0.96
					OPQ:	0.900

Exhibit 2.5 Inventory, Organizational Performance Quotient (example).

their indexed levels as well. The two averages at the bottom of the table are the indexed values for the organization being evaluated and the indexed BIC values. A simple division between averages (actual and BIC) provides the summary OPQ value. As mentioned previously, the BIC average value will most likely not be 1.0 because BIC is generally less than world-class level benchmarks. By using the simple equation (2.2), it is evident that the goal for the OPQ value is to be equal to or greater than one.

The most difficult part of creating an OPQ is the establishment of indexed tables for each of the system vitality traits. It will take research and modeling to construct the initial index table. With use and tuning over a series of uses, the indexed tables become more accurate and credible.

2.5 Communicate to the Organization

> Effective communication is 20% what you know and 80% how you feel about what you know.
>
> **Jim Rohn**

At this point, you are aware of the effort and energy required to undertake this journey. You are still reading because you know you can pass the hurdles necessary to move the organization into a new competitive position by LWL. Now, it is time for you to formulate your communication plan.

As the quote by Jim Rohn describes, so much of the effectiveness of communication is associated with the passion for the subject. This is extremely relevant in LWL. For you to get the organization engaged, you have to be engaged. An antiseptic or removed approach to this subject will not do. Your passion for learning, for application of that knowledge, and for change are the critical elements in the communication plan.

2.5.1 Complex Message Delivery

You started the transformation journey with basic activities of assessing the as-is conditions and environment. As you move through the transformation phases, the need for effective communication is rising quickly.

In Section 1.8, the need to tailor communications, in both content and delivery, was noted. The messages and specifics were based on the audience segment (Figure 1.10). The need to expand that concept is seen in Figure 2.7, and is accomplished using differing levels and speeds of porpoising communication tracks. Now, the communications need to align with the varying needs of the audience by phase of transformation. You can see that the degree of content, specificity, and proportion change over time even within the same audience group. This is largely due to the changing needs within the audience as they acquire skills, change roles and responsibilities, and evolve their styles of organizational learning.

In Figure 2.7, you will see an example of one communication thread represented by the porpoising line and nodes, down the audience vector in Phase 1 of the transformation, and conversely it looks at the porpoising of communication content down the phase of transformation vector for Kaizen teams. The porpoising is based on phase one through five as the organization matures, and by the types of audiences receiving communication.

By projecting forward, you can see multiple waves for each phase by audience or multiple waves by phase of transformation by audience. The graph would look complex if fully populated, but it is easier to describe when decomposed.

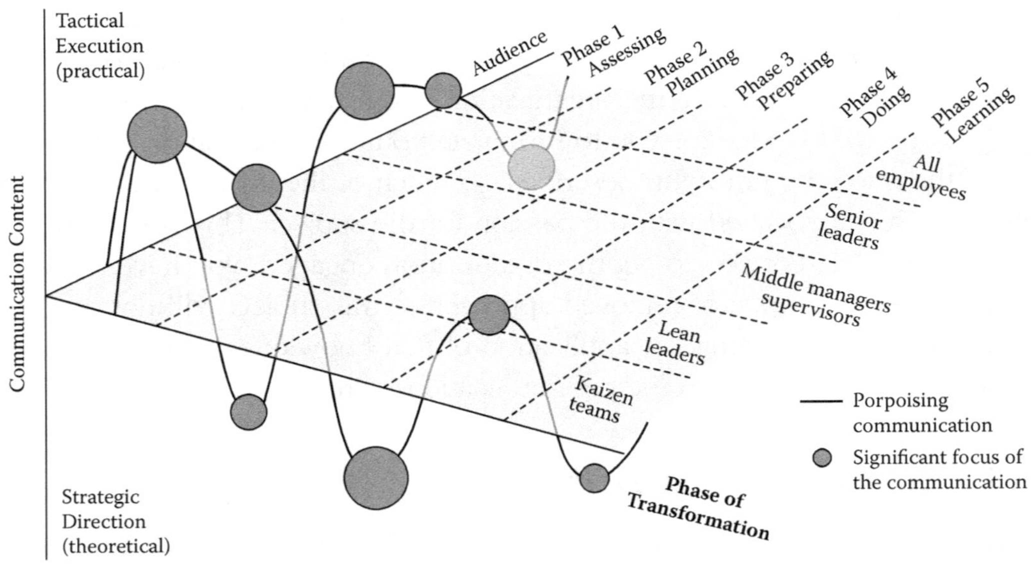

Figure 2.7 Proportion of communication content by audience segment and maturity.

2.5.2 Sharing the Transformation Vision

Earlier discussions demonstrated the need to use your personal cash to make this transformation. You cannot, and will not, make the investment necessary if you are not absolutely and passionately convinced of the direction. You will need to leverage your personal cash gained from all of the years of experience, hard work, and focus to make this journey successful. Unless you are passionate about the journey, you will not continue to invest through the long haul. Make no mistake at this point—LWL is not an impulse buy!

As you start to consider the macro level attributes of the end game, you have to come to grips with the fact that not everyone employed today will be part of the team tomorrow. Not everyone is associated with enterprises growing so quickly that doubling the efficiency could be made up in equivalent incremental sales. The reality is that some of the team will be left behind—some because they choose not to get on board with the new direction and others because the organization does not need everyone that it once did. This is a challenge for anyone who sees the potential. You have to believe that while you may not be able to guarantee employment for everyone, you can guarantee that every employee will be more valuable to this organization, or others, because of the effort they put into this evolution.

2.5.3 Setting the Course of Action

One of the toughest challenges will be to convince the organization that the risk of doing something, in this case LWL, is less than the risk of doing nothing. That said, there are risks, and this strategy like any strategy is not perfect or capable of covering all contingencies. That said, you will be developing your capabilities from the ground up. You will be investing in each employee with training, mentoring, and hands-on opportunities to apply their learnings. Growing this capability organically is one of the great advantages of LWL as a strategy. It cannot be bought; it is the culmination of a thousand steps by every employee.

Having a sense of the attributes of the organization available in the future, a personal passion for LWL and crossing the bridge reminds you that some members of the organization will be left behind if they cannot get on-board with the transformation. You can begin to focus on the actual plan to implement the transformation. You can start thinking of the sense of time it will take. The plan, and its timing, is very specific for each organization.

This journey has many similarities to a home renovation; the length of the project will be dependent on the amount of change required and the amount of money you have available. If the personal cash you bring to the effort is minimal, then plan for a very long transformation. If you have a high level of personal cash, then you can expect more effort and progress sooner. In either case, you are dealing with months, quarters, and years, not weeks and days.

There are three important components of the communication messages: sharing the as-is strengths and challenges, clarifying the gap between as-is in the end game, and crystallizing top-level action plans to guide the organization's transformation. Each of these will be briefly described:

- **Share As-Is Organization Strengths and Challenges**—Assessments of competencies and inventories of skills are done, and inputs have been integrated, analyzed, and distilled. From this work, you have a clear picture of how the organization is positioned with its strengths, gaps, and opportunities. You know where you have ideal behaviors and culture in place, where effective and efficient practices exist, and where your tools, techniques, and improvement methods are strong. Corollaries to these are that you know where process opportunities exist, where gaps in methods and practices have been identified, and ineffective behaviors and cultural misses reside. What is critical in the messaging is accuracy, honesty, and transparency. If these are missing, the organization will see right through the messages and the transformation will stop before it really gets started. Let the organization know what it does well, what it needs to stop doing, and what it needs to start doing. This sets the stage so you can increase the tempo and intensity of embedding Lean into the fabric of the organization.

- **Clarify the Gap between As-Is and the Vision and End Game—** As part of the messaging, gaps are important to understand; they define the size and type of void that needs to be filled or bridged. This gives an opportunity for the senior leader to clearly reinforce the vision and clarify what the end game looks like. This is a journey, one without an end but clearly with a direction, of course, at a rate of speed to get there. The organization has to be constantly refreshed and rejuvenated during the transformation. As the leader, this is Job 1. Re-reading the earlier quote by Jim Rohn, it talks about what you know and how you feel. Now is the time to let the organization know exactly what is in your head and your heart. It should be clear that you do not know all the answers at this point, but you have the passion and conviction to discover the exact right answers, and then transform them into reality uniquely tuned to the organization. Through your words and actions, you can excite the organization into breaking free from its status quo. If you truly believe transformation is required, make it known to the organization in no uncertain terms.
- **Crystallize High-Level Action Plans for the Transformation—** Supporting both of the first two components (as-is situation and identification of key gaps), articulation of a clear plan of action is important. This should be a high-level action plan, not a detailed one. Why not build a finely detailed plan? Using the words of Winston Churchill, former Prime Minister of Great Britain, "Plans are of little importance, but planning is essential." This implies that plans, while best laid, are frequently outdated as soon as they are written because of the fluid nature of the environment. However, the activity of planning forces alignment across the organization, cohesion between partners, and a shared vision by all.

As mentioned previously, it is important for everyone in the organization to know that you do not have all the answers. Environments change, customers' needs and expectations change, and the competitive landscape is vastly different from day to day. If you have a high-level plan that provides flexibility, adjustment, and adaptation to the changes, it will serve the organization and yourself well.

2.5.4 Personal Actions and Passion for Change

Socializing the macro concepts and benefits with your direct staff is also appropriate at this stage. You will also need to start thinking about where real change will take place.

- What is the embarkation point for the LWL journey?
- Which group, team, or individual will you have take on the first projects?
- How will you share the successes?
- How do you communicate the new direction?

These are some of the areas that you will address in the near term. Your passion will start to force you to talk about LWL. Do not be surprised to find the ideas being front and center of almost every conversation, arbitrary of the level of organizational maturity.

Key Messages in This Chapter

1. It is easier to recognize the failures of others before you see those of yourself.
2. Learning atrophy is a given condition as people and organizations age; of critical interest is organizational constraints—these are fixable (as opposed to unfixable biological ones).
3. Do you see the beauty of One-Touch Flow? You must be able to see and believe this is the right destination in your journey.
4. You need to grasp the concept and vision of Process Design Minimalism; it is an industrial art form that everyone is capable of mastering.
5. The Credibility Cash Index concept is foundational to your ability to move an organization into a transformation. You start with none and grow your bank account over time (with successes).
6. Assess your conviction to lead change; it is cathartic. Do it in earnest and honestly, and you will reap learnings manyfold your investment in time.
7. Avoid introduction of technology into a process or system laden with inefficient/ineffective steps; it will only make it harder to simplify in

the future and it creates process dust (techno-waste) that will never get cleaned up.

8. Understand the organization's Lean knowledge, practices, and commitment.
9. Understand the organization's capability to move toward (transformation) being a Learning Organization.
10. How locked in is the organization? It will be harder to change the longer it has been in a steady-state condition.
11. Subject Matter Experts from within the organization are the best sources of information and evaluation about the strengths and weaknesses.
12. You should be able to identify and assess the adaptability and degree of learning going on inside the organization and yourself.
13. Based on your conviction to lead this change you should be able to have a sense of your ability to accomplish the task.
14. Using the assessments there is an understanding of the organization's capabilities relative to learning, and scale of its inertia to learn or adapt.
15. Develop a sense of the challenges in communication that will result for attempting to deploy this strategy.

Challenge Actions

This section is about knowing what you have. It has both personal and organizational examination actions. The actions focus on intellectual honesty, introspection on your learning ability, and analysis of your belief system that you can and want to lead an organization's transformation into a Learning Organization. Choose action items that are important to you and can be accomplished in a week. The actions selected should total at least 50 points and come from at least two groupings.

Challenge Actions—Phase 1, Taking Stock	*Points*
1. Take 20 minutes and complete the simple assessment of your personal leadership and ability to lead a Learning Organization (*see* Leading a Learning Organization assessment instrument in the Appendix).	25
2. Take 15 minutes and complete the simple assessment of your personal leadership and your ability to lead Lean in your organization (*see* Lean Leadership assessment instrument in the Appendix).	20
3. Set up a 360-feedback approach to capture your managers', peers', and subordinates' rating on the Credibility Cash Index elements.	20

4. Use the scores, findings, and recommendations from the assessments and CCI to develop five or six action steps that you will use to begin changing your CCI.	15
5. At your next staff meeting, place on the agenda a topic that causes each member of staff to describe an event and recent learnings that resulted.	10
6. Build a list of personal work processes that you can improve when you have discretionary time. During the course of the book, you will pull items from the list to do work.	15
7. In 100 words write a description of your as-is leadership strengths, competencies, and behaviors with respect to continuous improvement. Save this until the end of the book for comparison.	10
8. Using a "T" format, label the left side as "drivers" and the right side as "restrainers." Now list the top five or six driving forces for why you want to adjust your leadership style, and on the right list the top five or six restraining forces that prevent you from making these style changes.	15
9. Make a list of potentially disruptive technologies and compare and contrast against your technology roadmap.	5
10. Can you drive to a minimalist culture? Pick two people who know you well, you trust for clear and honest feedback, and are outside where you work and ask them to describe your personal work style with regard to being simplistic or complex. Do the same with two people from inside work that you trust (and they trust you).	15
11. Using a short timeline, enroll your staff, supervisors, and select subject matter experts to complete the simple assessments of your organization.	25
12. Construct and measure your organization's OPQ.	20
13. Use the scores, findings, and recommendations from the assessments to develop an inventory of significant gaps that will be used during the following Planning the Transformation phase.	30
14. Conduct a series of brown bag discussions across the organization to capture a real understanding of customer focus and energy, process and system frustration, and employees' willingness to invest personally in being members of a Learning Organization.	15
15. Craft a list of organizational tattletales that you and your leadership team can use to see if the transformation is beginning to take hold and move the organization. They should be simple to see and central to the components of your transformation plan.	10

Chapter 3

Planning the Transformation

> A corporation is a living organism; it has to continue to shed its skin. Methods have to change. Focus has to change. Values have to change. The sum total of those changes is transformation.
>
> **Andrew S. Grove, 1936–present**
> *Engineer, Author, Business Leader*

Now that you have a lot of information in front of you, what do you do with it? You know the strengths and opportunities of the organization; you have a landscape of your personal strengths and opportunities, and a rough gauge of your personal CCI to guide you toward the amount and type of change you would like to achieve.

In this chapter, you will cover architecting and planning a transformation plan that your organization will use moving forward. It will target four principal actions:

- Designing the capability to learn as an organization
- Using the Voice of the Customer (VoC) as a focusing engine
- Leveraging Lean as a learning vehicle
- Embedding use of data as the root of process knowledge

Each of these receives an examination of what they are, why and how they are used, and why they are important to the organization's transformation toward achieving a sustainable competitive advantage.

Part of this chapter's content also uses two orthogonal challenges to those above that, as leaders, must also be overcome:

- Moving out of your comfort zone and becoming uncomfortable as you lead change in an organization, and
- Prioritizing the "tyranny of the urgent" (e.g., the "always present" need to respond to daily challenges, problems, and opportunities) in lieu of proportioned time allocations for both execution and improvement/process waste reduction.

In Figure 3.1, you move forward to the second phase of the journey where all of the data and information gathered during the assessment phase is combined with the leader's vision, direction, and speed to establish a transformation architecture and implementation plans.

The output of this architecture and planning phase is a plan, bounded by the vision, direction, and speed, and tuned by the identified strengths and opportunities of the leaders and organization. It will likely include:

- Expected behaviors (accentuation of existing ones and new ones)
- Style of Lean implementation
- Training requirements
- Defined deployment roadmap

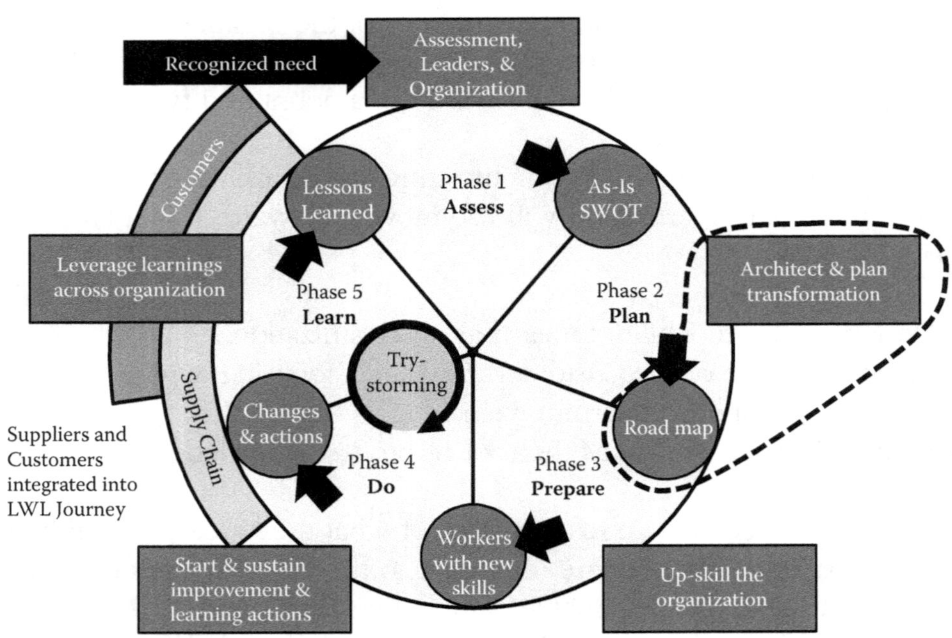

Figure 3.1 Architect and planning, phase 2—transformation model.

- Identification of implementation leaders
- Measures of transformation success and measures of impact to the organization

At the close of this chapter is a discussion about communication during this phase. The most critical points are modulation of message details, adjustment of content to audience, and frequency to ensure knowledge and awareness, without over-communicating into saturation.

With the information you have in hand from your personal and organizational assessments, you can begin to visualize and construct your vision, end game, and transformation plan for the organization. At this point, your ability to cross-link and compare findings and results, and searching for common denominators, key messages, and key themes, will give you the greatest baseline to begin constructing your plans.

3.1 Forming Your Personal Vision and Plan

> You have to pretend you're 100 percent sure. You have to take action; you can't hesitate or hedge your bets. Anything less will condemn your efforts to failure.
>
> **Andrew S. Grove**

As the leader of the transformation, its success or failure depends on your personal vision for its future state. In this section, there are four aspects for your consideration:

- visualizing the end game
- setting the direction
- personalizing the transformation
- walking the talk

Each of these is distinct and specific just to you. They will form the anchor of your conviction for change and your ability to communicate what you know in your heart and what you think needs to happen to make it real.

Earlier in this chapter, it was shown that you need to be directionally correct with your vision of where the strategy will lead the organization without being absolutely definitive in all aspects of the journey or end point. To accomplish this task while meeting the requirements of porpoising may seem like a contradiction. In fact, the requirements are very synergistic.

Understanding which attributes of the ideal state will help you to articulate what the work environment could look like in the future. Some ideas are outlined that could help to accomplish this shortly. Once you have these anecdotal references of the future, state what is that you are asking the organization to follow. Briefly, the following is a description of an end game using a sports analogy:

> An end game is the final competition or series of competitions that use a playbook of actions to compete and win a market or position, but without knowing exactly what that looks like (no end score is known, number of successful/unsuccessful plays executed, or even how the competition is positioned).

The basic strategy is to create a sustainable competitive advantage by installing LWL as part of the culture. What are some of the aspects that make this more meaningful to your organization? You have every right to expect the organization will be more efficient, have higher quality, be more responsive to external factors, and be better at NPI and time to market with solutions. Whatever the product or service that you are providing, think of the BIC examples of today and imagine what a better performing version of that would look like as it relates to your organization. The mental image of the future state that you create and share will help the organization understand where it is heading. The action plan and personal investment in the training and practices of LWL will get it started.

Now that the conceptual need for the vision of the future state has been addressed, how can you start to build that image? Here are a couple of truisms that apply to this strategy and can help you form this vision.

- The journey being described will unlock performance potential that is so large that you will have to use discretion in describing what you think will be possible.
- Experience supports the position that most unstreamlined processes contain a minimum of 80% of the energy or activity associated with them (later defined as system entropy) as non-value added.

What do these mean for you? A few examples of how this translates to a "de-tuned" set of expectations are:

- Double the output with the same amount or less resource
- An order of magnitude improvement in process time or throughput
- 50% improvement in development cycle, product/process/service launch, or time to market

By constructing an overview of the potential that the future holds, you can begin to engage the minds of the long-term thinkers in the organization. Your description of the future to the more pragmatic elements of the organization will have to have much more specific data-based content.

3.1.1 You Have to Buy-In

Having set a very high expectation level for the comment from you as a leader to be successful is in fact dictating that this solution is only an option for the very few (or the very brave?). No! It is only conditioning you to appreciate the amount of commitment and effort required to move the organization utilizing LWL.

The journey will be long and have many curves; it will not be straight but rather characterized by spurts of progress offset by setbacks. It is against the organizational obstacles of inertia from success, lack of external stressors, and the characteristic of a very small rate of change at the onset of this journey that make LWL a challenging strategy to implement. You have to believe that you are on the right path and remain committed to the goal. If you have this level of buy-in, this strategy is the one that enables an organization to be successful in front of almost any challenge.

3.1.2 Use Your Passion and Courage to Lead the Organization Transformation

> Whatever course you decide upon, there is always someone to tell you that you are wrong. There are always difficulties arising which tempt you to believe that your critics are right. To map out a course of action and follow it to an end requires courage.
>
> **Ralph Waldo Emerson**

By now, the gauntlet has been thrown. Do you have what it takes to lead this transformation? Before starting to prepare your readiness, let us flush out a few more criteria of success. The need to be able to porpoise at every level of dialog to help people understand the direction has already been covered. Added to that was the concept of adding energy, or creating stress, required to overcome the lack of external threats and the inertia from being successful.

Beyond constructing the roadmap to execute this strategy, what do you need to get started? Passion! Your conviction that this long-term journey will enable the organization in a way that cannot be replicated with money, outside consultants, or short-term effort is what is needed to succeed. You absolutely have to believe that the effort required is not only part of the barriers for others to replicate this strategy, but also is a vital part of the transformation inside the organization. It is your unflappable confidence in the direction and need for change that will provide team members the guiding light if and when they are not clear on where they are headed.

Before beginning you need to internalize three leadership capabilities that you must be able to demonstrate in order to implement LWL:

- **Ability to porpoise**—The macro- and micro-level view of the journey to allow you to porpoise with the staff in your communications to them.
- **Energy and stamina for the journey**—This will be a long journey that will require you to put a disproportionate amount of stress and energy into the organization to start LWL.
- **Passion to transform**—This is not an academic exercise working with theoretical models. You are asking people to change the way they think at work. You are asking many of the team to re-engage their ability to learn and it will be met with resistance and skepticism. Your strength of conviction and passion will be tried!

3.1.3 Visualize the End Game

Today, you and your organization are performing well in many to most areas. Sustainability during market changes, emerging competitive organizations, ever-changing regulatory and government requirements and constraints, and new disruptive technologies are always at your doorstep. Creating and sustaining a competitive advantage is critical to long-term success. The foremost action that helps the organization stay focused is having and being able to communicate clearly an ever-changing end game.

The manager accepts the status quo; the leader challenges it.

Warren G. Bennis

Questions quickly emerge as you form the end game. Many are likely to sound like:

- What does that look like for you?
- For the organization?
- For your customers and supply chain partners?
- For the key stakeholders?
- How quickly can you change it as the environment changes?

These and other questions must be used to help form your personal vision of the end game. Whatever the method used to help, it must provide you with four basics elements of your end game vision:

- Direction—clarity regarding where are you heading
- Course—flexible paths you will use to get there
- Speed—knowing how fast change is needed or absorbed
- Relateability—every employee see themselves in the end game

This last one, relateability, is a key element of making the vision real for the organizational. You can synthesize and craft a complex end game, but it will be of no value to the organization because no one can understand it, they cannot remember it, and they will almost never be able to communicate it to others when asked. In fact, it will become a moniker of an approach that is too complex, too burdensome, and too difficult—all pointers to yet another initiative of the month.

Make the vision clear, crisp, and memorable. It should be concise in its scope and measureable in its outcome. Employees will remember what it is, what it means to them, and what it means to customers and suppliers.

There is no perfect plan—never was and never could be. There are too many potholes, alligators, and trip-wires in today's environment to craft a perfect plan. With that in mind, you still need to create a plan that is concise, actionable, and measureable, and something that incites passion and energy in the organization.

As you build your vision, keep its constructs simple and with a big picture in mind. Usually, there is a strong pull to tap into the consulting and external resource market to get help. This is one of the few times where using the expertise of others is a detriment. This must be something you feel all the way to your core; you must have your fingerprints all over this guidance to the organization. Use consultants for many other things, but not this. Consultants do well with tools, training, and process methods, but should be used sparingly when it comes to what to work on, when to work on it, and who should work on it.

3.1.4 Personalize the Transformation

What will make this vision of the end game real for the organization? Simple—the organization must see your belief and passion in what it must look like, how you will lead the actions of transformation, and how you will be holding yourself accountable to ensure the inertia of the organization does not roll over the will to effectively change.

The test will be how you tell the story of what you see and believe for the organization. Your employees and leadership team are sharp; they can read you like a book and will know immediately if you are genuine or if this is a faddish change attempt. One way you win them over in a heartbeat, the other way you will get grudging compliance and the energy required to change will be far greater than what you have to offer. Keep the story simple; make it a "one floor elevator" speech:

- Simple concepts
- Simple terms
- Easy relevance to the mission
- Identifiable and relatable for all employees to see how they can be involved

The simple fact is that in order to be successful, you need the entire organization to understand what it is about, what direction the transformation is heading, and how they need to align, support, and enable the change tasks coming toward them.

Of the many behaviors and skills that you need to demonstrate for the organization, there is one that needs very clear and frequent demonstration in daily work. It is called One-Touch Flow (OTF) and is referred to as Kaizen eyes by Lean leader Taiichi Ohno in his book, *Toyota Production System* (Ohno, 1988), and seeing the flow by James Womack and Dan Jones in their book, *Seeing the Whole Value Stream* (Jones & Womack, 2011). OTF will be explored in greater detail in Chapter 5, but for the moment it is briefly described as the ability to observe as-is process execution in its most discrete actions, rapidly identify waste, ask pointed questions to uncover why something is touched more than once, and drive waste elimination.

OTF is defined as:

> The concept of touching any task, action, or deliverable just once in the value transformation chain.

The natural fears of job or position loss in an organization when changes happen has been alluded to. Starting now, you must be able to help employees see and understand that elimination of waste does not mean elimination of their job. It does mean elimination of wasteful work and that they need to be capable of performing higher value-add work. This may require them to acquire new skills or begin applying their existing skills in new ways.

Your responsibility is to make sure you have fast and ready answers for questions such as:

- How would resources be characterized if you free up capacity by getting back 80% of your resources as free (typical level of entropy in knowledge work)?
- How will you use the newly liberated resources to address the predefined issues, opportunities and challenges (remember—the aim is to keep organizational fear in-check that people will not be streamlining themselves out of a job)?
- What does the organization look like with
  - Large number of small changes versus one large one?
  - An increased number of functions involved versus layered hierarchy?

In the next section, we will be talking about building your personal action plan. Let us recognize that effective organizational changes start and are led from the top. Employees model after those at the top as a reflection of leadership, respect, and alignment. That is why your personal action plan is critical to the kickoff of the transformation work. This action plan must be reflective of your conviction, your vision, and your passion to make the changes real.

3.1.5 Build Your Action Plan

You now have a lot of information (several self-assessments and CCI inventory) that you can use to start changing the organization, beginning inside your office. The following will guide you through the "rack, stack, and select" exercise of identified opportunities, prioritization, and selection of the opportunities most important right now and building a personal action plan that starts the transformation journey with you.

Necessary and Sufficient Test—The concept of necessary and sufficient tests for the architectural design (or any other topical subject) is based on a two-test qualification schema. First, assemble a list of actions, tasks, improvements, and measures (or anything else of importance to the subject). Then arrange it in order of impact to the subject. Using this ordered list, start at the top and ask two questions of each listed item:

- If we do this [listed action], is this *necessary* to succeed? While some listed actions are important, they may not be necessary for success.
- If we do this item, plus all the other [listed actions] marked *necessary*, would they collectively be *sufficient* to assure success? When the answer moves from "no" to "yes," stop. You have just identified exactly what listed items are both necessary and, when collectively done, sufficient for the topical subject to be successful

Rack, Stack, and Select for Personal Action Plan—You have multiple lists of opportunities from three self-assessments you need to integrate. A simple rack, stack, and select activity will enable you to integrate the various outputs (racking), align and combine like type opportunities into a single item, and then rank them (stacking) in a priority according to a set of needs or criteria. By using simple techniques that include establishing a few attributes (three to five) and a basic scoring system, prioritization and selection occur quickly.

It is time to develop a plan for personal growth centered on organizational learning and Lean skills, methods, and practices. Create a quick scoring table using the outputs of the assessments. It should have three simple sections:

- Strengths (do more of these actions)
- Opportunities (new needs for specific skills and practices)
- Weaknesses (stop doing these actions)

In each section, place specific, descriptive language-defining actions identified from the assessment outputs.

3.1.6 Walk the Talk

Now is the time to take your vision, belief, and passion to the organization. Throughout the coming chapters there will be discussion about the change architecture and plans, preparing the organization by deploying knowledge and processes, putting the knowledge to work by creating value for the customers, employees, and stakeholders, and then leveraging the daily learnings into a useful pool of knowledge-based lessons applied across many parts of the organization. A few quick thoughts on how you can get your vision and passion more visible across your management team and the organization:

- Hit the floor. Talk to employees as partners in the coming journey. Enroll them, and you will likely find many have knowledge or background experience and are more than eager to advance the vision across the organization.
- Become a consummate communicator. You need to build a small library of ad-lib lines, make them well rehearsed, and use them often as you intersect employees across the organization and engage with improvement teams.
- Engage the layers of your management team. Most of them will quickly jump onto the vision and begin carrying it forward. A word of caution: While they are eager, they may not be aligned or even have the common language, knowledge, and sound bites needed to align and lead their organization. Bring them in, but hold them at bay for a short period while the architecture and deployment plans are resolved and refined.

Your employees will look to you as their beacon in the fog of change of where you are about to lead them. The fog of change (much like the fog of war) is the result of overload—too many things are happening at the same time, and their perceptions of what they see, hear, and feel are not keeping pace with each other. The impact is clouded thinking and judgment. This is normal, but needs to be handled so as not to lose them right out of the gate. Traditionally, organizations will exhibit signs of uncoordinated direction, misinformation, and confusion in actions that would not normally be present.

3.2 Changing Organizational Culture

The formula to assess changing the culture:
$H \times V \times F > R$
where:
H = Hatred for the current reality
V = Vision of the ideal state
F = Courage to take the first steps (take a risk)
R = Natural resistance to change

Jamie Flinchbaugh and Andy Carlino

Changing culture requires diligence, effort, and focus. In the article, "Cultural Change That Sticks," in a recent edition of *Harvard Organization Review* (Katzenbach, Steffen, & Kronley, 2012-90[7/8]), the authors summarized five factors to pay attention to as you try to change or transform a culture. They are presented in Table 3.1. They range from matching strategy and culture, to measuring and monitoring cultural evolution. As the senior leader, you are initiating and driving the transformation and these five factors must be etched into your personal behaviors. As you develop your personal improvement plan using the many various inputs from the assessment tools and CCI inventory, there are points of intersection that emerge and reinforce the need to personally emulate these behaviors in your own, tailored way.

The importance of these factors becomes evident when you internalize how it will affect changes to your daily work habits. Those around you—your peers and leadership team—should be able to see evidence of growth and transformation in your personal work styles.

By understanding the five factors of changing a culture, you are ready to move into the architecture and construction of the transformation plan. In the next section, you have the opportunity to tailor what you have heard and what you know into a plan. Your challenge as leader is to prevent the plan from dying under its own weight—it must be specific yet allow for changes as you progress, agile in its ability to be changed quickly, relevant to the organization, and right sized in the resources required to make it reality.

To sum all of this up simply, don't over-think the solution.

Table 3.1 Five Factors for Changing a Culture

Factor	*Actions*
Match strategy and culture	• Focus on leveraging the best elements of existing culture with desired strategies • Remember, culture trumps strategy every time—be mindful of avoiding strategies in competing directions
Focus on a few critical shifts in behaviors	• Observe prevalent behaviors today, compare with desired behaviors if the organization is acting at its best • Ask some of the best performers in the organization about what parts of the culture affect worker behaviors—good and bad • Select the vital few with the greatest impact to work on
Honor the strengths of your existing culture	• Don't dwell on the negative traits • Acknowledge the positive cultural assets to reinforce a shared evolution of the culture • Search, find, and leverage pockets of excellence—units of the organization already aligned to the changing culture
Integrate formal and informal interventions	• Balance formal interventions (e.g., rules, metrics, incentives) with informal interventions (e.g., communities of interest, visible use of the best performers, leverage from pockets of excellence)
Measure and monitor cultural evolution	Focus on four areas: • Organization performance • Critical behaviors • Milestones • Underlying beliefs, feelings, and mind-sets

3.3 Changing Organizational Momentum

Resistance is thought transformed into feeling. Change the thought that creates the resistance, and there is no more resistance.

Robert Conklin

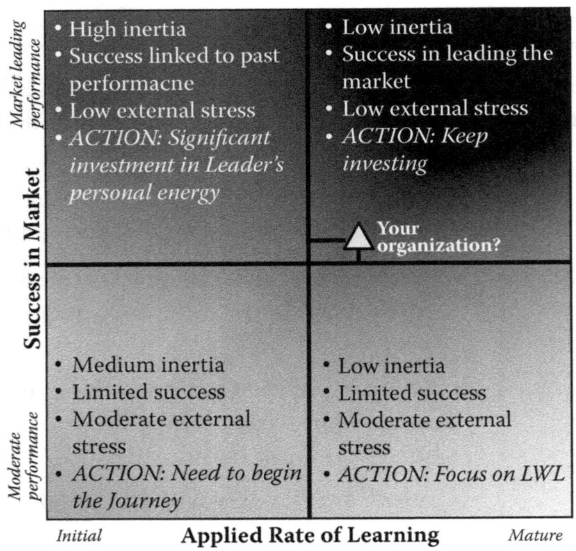

Figure 3.2 Organizational momentum versus organizational inertia.

This section is all about changing an organization's momentum. It draws analogies between organizational dynamics and the laws of physical motion. The quote by Robert Conklin strikes at the heart of these parallels. The concept is that resistance is the result of people's thoughts (fears related to change) being transformed into feelings (their natural resistance to change). By changing people's thoughts through knowledge, environmental receptivity, and consistent leadership expectations, the results diminish an organization's resistance to change.

In order to set the stage for the analogy between organizational dynamics and the law of physical motion, you need to explore the relationship and results that impact changing momentum. Figure 3.2 uses two axes to create a scalable field: Success in Market and Rate of Organizational Change. The scales for each one are linear, and you can see representations of minimum and maximum rates based on the size of the balls at each end of the scales. A quick discussion for each plus the inertia field follows:

Applied Rate of Learning—The horizontal axis denotes the rate of change an organization shows as evidence for its evolutions of change over recent history.

Success in Market—The vertical axis represents a good performing organization with a stable market presence (remember, our incoming premise for this book is that your organization is already a good performing organization). At the top of the scale, you see market-leading performance with the organization demonstrating a commanding presence in the market versus its competitors.

Inertia Field—This is graduated from upper right (low organizational inertia) to lower left (high organizational inertia). The scale is exponentially increasing as it moves from upper right to lower left.

So, how would you know how to grade your rate of organizational change? The following tattletales are proposed as some of the evidence to see that cycle of evolution has occurred:

- A significant change in the market served, or the success criteria for a given market has significantly changed.
- The integration of an entity (an M&A acquisition) that represented a sizable proportion of the host organization.
- Demonstrated evidence of a fundamental shift in the organization's culture, whether as a result of leadership direction, changed environmental conditions, or induced by competitive stress.

In physics, inertia is described as resistance to the change in state of motion in a body. In this analogy, it parallels the resistance of the organization to change from its current operating norms to an alternate set of behaviors. Equation (3.1) represents momentum change as a result of the integration of organizational inertia divided by the credibility cash invested to make the change.

$$\textit{organizational momentum} = \int \frac{\textit{organizational inertia}}{\textit{CCI invested}} \tag{3.1}$$

Using the analogy described, physics declares that momentum is equal to mass times velocity. Organizationally, if we equate mass using the business systems required to sustain success in the market and velocity as the rate of change of evolutions in the organization, you can quickly calculate a measure of momentum by using the scales of success in the market and rate of organizational change.

First, the organization's momentum begins to rise slowly, because of the stress for change. Second, organizational performance turns down slightly; this is attributed to the disruption of the status quo, the introduction of stress, and the natural human resistance to change. From the midpoint on in this representation, you see the rate of organizational momentum again change more rapidly. Correspondingly, the downturn in organizational performance reverses itself and begins an upward climb. As you proceed to the right side of the chart, organizational inertia is beginning to take off. The rate of organizational performance is beginning to rise dramatically. At some point in the maturity of the organization, these two lines cross. This event horizon is when the organization is in self-sustaining mode.

Why is all this important? Because it is relevant to the amount of energy required to initiate and sustain your organization's transformation journey. Let us recognize that it is human nature, in the beginning of a transformation, to underestimate the amount of change required across the organization. The likelihood is high that you will be too kind in your assessment and will underestimate the distance between where you are today and where you want to be (your end game). When this happens, you will be unprepared to face the challenge, the personal investment of your credibility required for the change, and the amount of time it will take.

Setting all jokes aside, a quick rule of thumb is to take your original estimate, change the unit of measure to the next higher unit (e.g., day to week, week to month, and so on), and then double it. Here is a quick exercise: write the original estimate and the rule of thumb adjusted estimate on a piece of paper, seal it in an envelope, and put it in your desk drawer. When you reach the significant milestone at the end of the original schedule (likely to have been adjusted several times since the original one), take a moment, open the envelope, and compare, honestly, where you are at in the journey versus what it is going to take to fully complete the transformation. Then, look at the projection on the piece of paper; you will be surprised at how closely they align.

3.4 Design the Transformation to a Lean Learning Organization

> I am enthusiastic over humanity's extraordinary and sometimes very timely ingenuity. If you are in a shipwreck and all the boats are gone, a piano top buoyant enough to keep you afloat that comes along makes a fortuitous life preserver. But this is not to say that the best way to design a life preserver is in the form of a piano top. I think that we are clinging to a great many piano tops in accepting yesterday's fortuitous contriving as constituting the only means for solving a given problem.
>
> **Richard Buckminster Fuller**

A typical approach to establishing a plan is to articulate the macro level concepts to senior management and let them add granularity as required through the hierarchy to the extended organization. What makes that approach inappropriate for an LWL strategy? In this example, there is reliance on many layers of management taking big-picture concepts and further refining them into a meaningful plan with their respective teams. The act of accomplishing this task is predicated on years of cultural, social, and industrial learnings, vernacular, and customs. With LWL, you do not have these basic building blocks to leverage, so the plan that you construct will have to have to be much more sensitive to the needs of a large audience.

The plan must be made so the content resonates with all. It must span the organization to include all types of employees (top to bottom), all fields of work type (transactional, design, manufacturing, service, and support), and all years of experience This is different from any other plan that you have probably created because this is not one that you can easily delegate to your staff to implement. Here again, you see the level of commitment necessary to make LWL a successful strategy for your organization.

Let us explore two elements required for this plan:

1. **Planning horizon:** This is a journey and the near term result will be small. As you consider the changes that you will want to see in a given horizon, apply this simple rule: the time it will take to approach this

goal will be 2× what you currently believe the case to be, and at the same time you will achieve only half of the impact expected! This will be a long process, so plan accordingly.

2. **Level of granularity required:** This is another dichotomy that you face. The need to articulate meaningful elements of the plan at various levels in the organization dictates the need for specific details. In contrast, the further away you are from the process, the less you know about how the process works. This apparent dilemma is solved by two considerations:
   a) **Target the audience:** Ensure that your ad-lib lines can be well rehearsed. You will need to have specific near term problems to unleash on the organization. Using your knowledge of the waste that exists, the benefits in simplification and the improvement in flow or quality will be sufficient to excite some of the staff to fix the problem using Lean.
   There is the concept of "low hanging" fruit in your organization, which can be seen in Figure 5.5. By focusing on the obvious, you will encourage the team to challenge the status quo and to take matters into their own hands to fix those items in their immediate control. By targeting these activities, you will help achieve some early wins, building your credibility and avoiding the knowledge trap of developing the solution rather than helping the team to identify an opportunity for improvement. If you are having trouble identifying what they could work on, simply ask a few people discretely what is the thing that annoys them the most about their job. Using the "5 Whys" you can construct a project for them to work on that is within their span of control and will make their job better.
   b) **Keep the plan simple:** The only true indicator of success that you have will be evidence of an increase in the rate of improvement. Therefore, do not construct a plan of infinite detail but one that allows the organization to evolve it at the rate of development that it can tolerate. This is where you must simply provide the appropriate directional input and let the initial successes start to flow. Using specific targeted engagements as outlined in the first item and very general directional input into the broader organization will help people absorb the new direction.

With these two basic elements of the plan described, you can further refine some of the other elements in the plan. In any change process, there is a standard adoption curve. In LWL, you have to be conscious of the two ends of the adoption curve (see Figure 1.6).

You must aggressively endorse the innovators/explorers and early adopters, giving them access to training; publicize their success not to illustrate the "organization impact" (remember, these may be small or immeasurable changes at this time), but as evidence that change can happen and it can produce a tangible benefit to the employee.

For those on the other end of the adoption curve, you need to place clear expectations on these change resisters/laggards. Here, the message is simple—change is inevitable and they need to make a decision on how they will engage, or they will be rendered obsolete. Their resistance may be caused by fear of the unknown, doubt in their own abilities, concern with losing knowledge power they currently enjoy, and so on. Helping them get on board with the plan or prepare to quickly leave the organization is critical to the future success of LWL.

The other point worth adding more color to is identifying the areas you can challenge or focus the team for improvement. Here, it is strongly recommend looking outside the organization for examples of BIC performance. Your personal experiences in returning an item to a high-end store or ordering something online can give you concrete examples of efficient processes.

- Are your internal processes easy to use?
- Do your customers feel as valuable after dealing with your organization on a defective product?

There are countless ways to take external data points and do simple gross comparisons on the experience. This is not meant to be an exhaustive study, but rather a simple illustration of where there may be low hanging fruit.

In Figure 3.3, there are two views presented—most frequently the internal view and less frequently the external view. Successful organizations have well-developed techniques to examine processes internally that are performing well, comparing them with BIC processes, and capturing learnings from the differences. In later chapters, there will be more time spent on benchmarking the processes and results, and how they should impact goals and objectives.

Less-developed organizations traditionally have fewer external views of BIC organizations in their particular industry, or world-class organizations regardless of their industry or mission deliverables.

It is vitally important that organizations seeking to build and maintain a sustainable competitive advantage effectively balance the need between

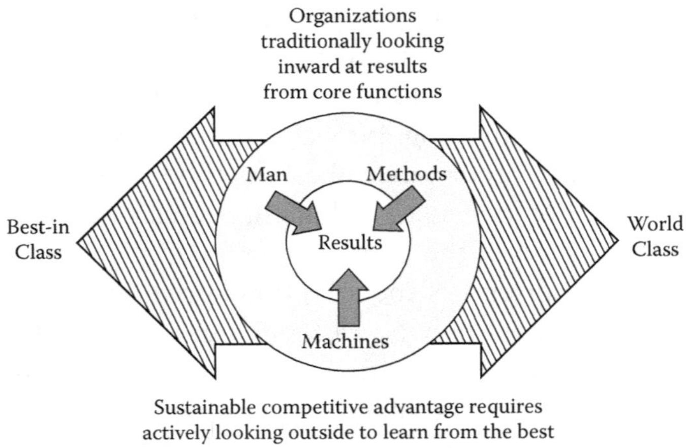

Figure 3.3 Learning from external views of best practices.

internal and external views. What is not always easy, but still very exciting, is to examine other external organizations to see what they are doing, how they are doing it, and learn new methods.

The challenge is to constantly view the return on investment in time and resources required to capture the knowledge and convert it into actions. Use caution because many benchmark activities, while fruitful and bearing significant information, become redundant for new or breakthrough knowledge. Always engage benchmark partners using pragmatic criteria to determine the payback value to the organization from the information acquired versus the investment of resources needed to acquire the information.

As part of the architecture for the transformation, organizational change models must be found that best fit an organization's existing culture and behaviors. In Table 3.2, three examples of change models have strong affinity and alignment between emotional, behavioral, and tactical actions required to institutionalize an organization's transformation.

The first model developed by Kurt Z. Lewin (1947) uses three simple transformation stages: unfreeze, change, and freeze. These stages are simple to understand, have logical inflow, and are easy to communicate to the organization. Lewin's model is frequently referenced to understand the emotional changes required during organizational change.

The second model by Jeffrey M. Hiatt (2006) is built on five behaviors: Awareness, Desire, Knowledge, Ability, and Reinforcement (ADKAR). Understanding the behaviors allows you to begin the translation from emotion into action using behaviors as a way to coordinate and reinforce change action.

The third model is by John P. Kotter (1996) and contains eight actions that include: change urgency, coalition for change, vision of ideal communication, empower to act, get quick wins, consolidating wins, and institutionalization. Each of these steps is easily converted into a series of tactical activities, with each activity aligned through the behaviors up to reinforcing of movement through emotional states.

While there are many change models available, these three are proposed to have significant impact to the effort of transformation for LWL organizations. In Figure 3.4, the three individual models are woven, or overlaid, into three coordinated rings with supporting sections aligned. Working from the outside in:

Outer ring—Start by understanding the three emotional states of the first ring, what is required to move from one to the next one, and what it takes to move people through the complete set of emotions. The Lewin model is used as the starting baseline. It is positioned as the outermost circle, which provides a coarse but easily understandable framework, and it has the ability to enclose and align with the Kotter and Hiatt models effectively.

Middle ring—The second level in the transformation model focuses above the emotional stages, and addresses a set of five behaviors characterized in Hiatt's ADKAR model. The five organizational behaviors provide deconstruction of a change method that forms a desired cultural expression of how the organization's employees treat each other, the organization as a whole, and the supply chain partners as integral members of the organization's value chain.

Inner ring—Finally, encased by Hiatt's ADKAR framework is Kotter's model represented as the third ring. It is composed of eight tactical actions that further refine behaviors into actions used to make the transformation occur.

When viewed as a coordinated triad of change models, the beginning starts with the need to emotionally unfreeze an organization. You have to recognize that there is both an awareness of and urgency for the need for change, and that there is a need to have a collaborated (common) vision and the ability to communicate the vision to align everyone in the organization.

In the emotional change phase, you must have both knowledge and the ability to make the change. Implied within that is empowerment of employees, giving them the capability to make quick wins and consolidating the

Table 3.2 Comparison of Three Organizational Change Models

Change Models			
Model	*Elements*	*Strengths*	*Weaknesses*
Unfreeze-Change-Freeze Model (Lewin)	• Grounded in changes to emotional transformations • Simple three-stage model • Easy to communicate and get organization behind it	• Well understood and practiced • Three simple stages • Easy to communicate • Logical constructs	• Simplicity overwhelms effective implementation • The model is more complex than the simple three stages

ADKAR Model (Hiatt)	• Grounded in changes to organizational behaviors
	  • Awareness
	  • Desire
	  • Knowledge
	  • Ability
	  • Reinforcement
		• Intuitive stages
		• Aligns with logical behavior-change approaches
		• Stages have defined exit criteria
			• Operates on organization's deep-rooted behaviors, which may require additional effort to change

8-Step Model (Kotter)	• Grounded in tactical actions: 1. Change urgency 2. Coalition for change 3. Vision of ideal 4. Communication 5. Empower to act 6. Get quick wins 7. Consolidating wins 8. Institutionalization	• Actions follow logical flow • Easy association with continuous process improvement cycle • Steps have discrete actions and results • Tactical orientation drives measurable results	• Eight steps requires additional implementation planning and resources • While logical, speed of progression through steps may hinder faster changes

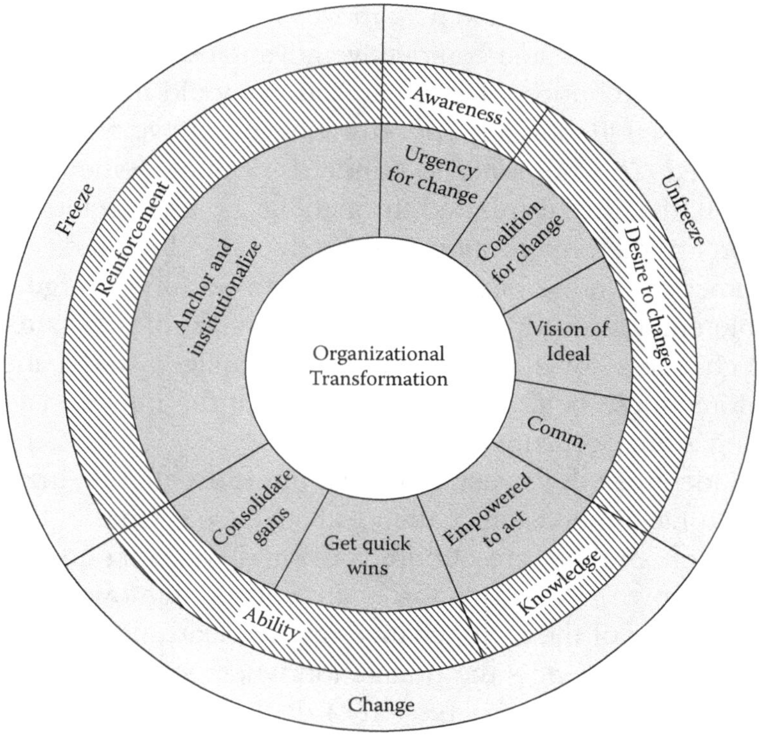

Figure 3.4 Transformation model interlacing.

gains. In this phase, training and the institutionalizing of Lean knowledge into the organization is established. In addition, this is where enrolling change leaders and the Lean leaders who have said that they can help drive and engage the rest of the organization is discussed.

Last, in the freeze phase of the model you see at the behavior level, reinforcement of changes that have taken place and what is needed to support tactical activities that provide anchoring and institutionalization. Typically, this is where policies are rewritten or created, new procedures are written or old ones modified, and daily behaviors of all employees are reinforced and recognized through active feedback and formal reviews.

3.4.1 You Can't Buy It—You Have to Build It

Throughout this book, we focus on the need to create hundreds of small, localized improvements, with each of these improvements owned by members at all levels of the organization, from front-line employees to senior level leaders.

The vision of the organization as it moves forward and matures is in order to provide a sustainable competitive advantage. This means that it must become change-hungry. The organization should be advancing to a point where it is self-identifying opportunities to remove waste and become more streamlined. This translates to hundreds of improvement actions, small fires if you will, that are distributed throughout the organization, at all levels, and in all parts of the organization. Ultimately, these hundreds of fires need to merge, sometime on the planning horizon, to become a large cauldron of changeable employees, driven by their intolerance of the status quo and the need to challenge for improvement. Their managers and leaders should openly reinforce these behaviors as evidence that the right behaviors and cultural changes are occurring.

Figure 3.5 identifies the sweet spot for a transformation being described. At the top, you have the vision of the ideal state (sometime in the future), and from the bottom there are hundreds of small, burning improvements thriving on their own. These two forces are converging toward the middle, becoming indicative of the ideal, change-hungry, learning organization.

Supporting this concept is the orientation where starting small and then growing large is the proper progression of deployment. This means that the transformation should be having thousands of small quick wins, the small improvement efforts by small teams and individuals, which yield marked improvements that are impactful to the team or the person doing the

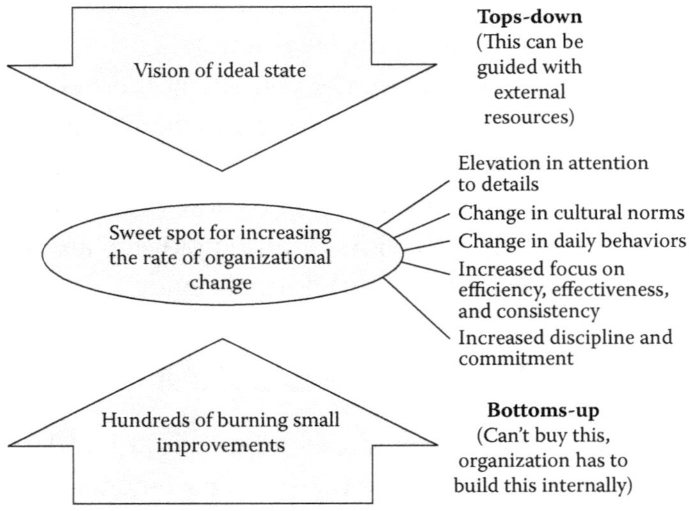

Figure 3.5 Top-down/bottom-up sweet spot for organizational change.

improvement, but may be difficult to quantify the impact to the organization at the macro level.

The advantage of starting small is that it prevents a Big Bang approach that traditionally turns most organizations against the change effort. Almost every organization has experienced the introduction of some type of Big Bang approach at some time in its history. Only a very small handful of these ever have been successful. The start small and grow approach avoids the appearance of an unwieldy, large program, a significant initiative that incites an initiative of the day response in the organization.

Architecture for the transformation has many parallels to the architecture of a new product, a new building, or a new service. It must be able to provide enough information and guidance and answer these questions:

- What is the Should Be state at some point in the future?
- Do you have a starting as-is baseline?
- What are you going to change?
- How will you change what is required?
- Who is going to change it?
- When will the Should Be state be evident, and what will the evidence be?
- What are the impacts to the organization, suppliers, and customers?
- How much will it cost to make the changes, and what priority is established for investment?
- What changes are expected with people's roles, accountabilities, and authorities?
- How will change actions be measured and reported (act of changing)?
- How will impact results be measured and reported?
- What tools will be used?

Use consultants if you must; use them to keep you grounded and to sharpen your skills, but avoid using them to take over the directions and construction. You have to grow it yourself. Let the subject matter experts (SMEs) pitch in and craft solutions that will work—they know the most accurate way things are running.

See the future and build the path to get there. Create general timelines; do not be too mired down in having precise schedules at the macro level. They are sure to be scuttled as organizational ebbs and flows occur.

In his book, *Quality is Free*, the author Phil Crosby (1979) used the premise that investing in improvement yields more than what you put in. In today's management philosophies, there are two camps: those who believe this is true, and those who do not. If you agree that today's processes and systems in most organizations contain 80% waste and you have the ability to remove half of that waste through modest investment in training, tools, and dedication of time, then Crosby's premise stands.

Are you willing to agree this is true and invest the time and resources to optimize the processes and systems? By now the answer should be a resounding yes. Crosby stopped short though—his focus kept him at a perspective that did not include the full impact of the environment, the market, disruptive technologies, and a changing customer landscape.

3.4.2 Integrating Your Supply Chain into the Journey

Thus far, we have been focusing on the development and transformation plan for your own organization. It is time to expand the scope and include supply chain partners as integral elements in your overall journey toward achieving a sustainable competitive advantage. Their inclusion must contain many of the same elements that you are expecting for your own organization, but phased in appropriately.

In Figure 3.6, you see that suppliers and sub-tier suppliers are not included in Wave 1 deployment. The need is first to have your own organization deployed and transformations started, and then expect similar transformations in your supply chain partners. Wave 1 is for your organization, and Wave 2 and beyond is where you extend your reach and expectations into your supply stream.

In Wave 2, the same kind of transformation elements that are contained in Wave 1 deployment for your organization need to be planned and communicated with adjustments according to each supplier so that they can

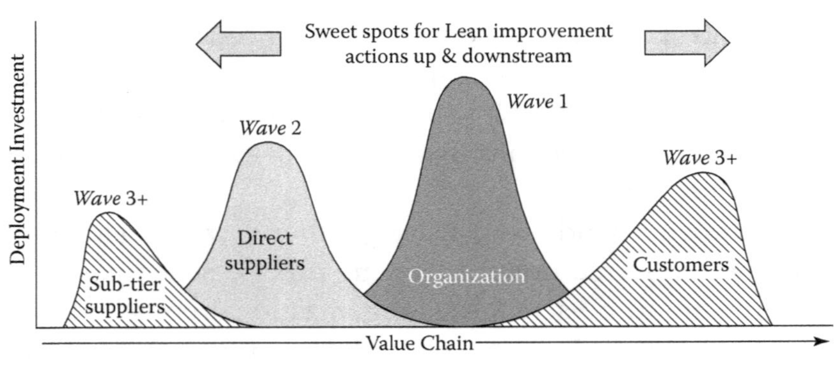

Figure 3.6 Integrating the supply chain and customers into transformation waves.

make the journey with you. These adjustments must take into account their size, maturity, the representative percentage of their business that your organization comprises, and the overall culture and behaviors that are part of their organizational fabric.

Do not expect your co-workers to exhibit complete eagerness to join the journey with you. It would be similar to your customers stating changed expectations in the core functions of your organization. In the design of Wave 2 transformation plans, include supply chain partners, considerations for recognition of overall performance, sharing of the benefits of mutual process minimalism, rates of Lean adoption, and improved profitability. Part of the transformation plans may also include the concept of supply chain partners segmenting their workforce so that your organization can selectively train, recognize, and integrate select members of its workforce to share the journey and the successes over time. It is like building an elite workforce of Lean practitioners in your supply chain partner's organization.

3.5 Lean Becomes the Learning Vehicle

> Waste is worse than loss. The time is coming when every person who lays claim to ability will keep the question of waste before him constantly. The scope of thrift is limitless.
>
> **Thomas A. Edison**

Let us openly recognize that there is a wide range of paths that an organization can use to guide its continuous improvement culture. Some of these are regimented and require precise methods, while others are very broad and non-specific in their expectations of systems and processes to meet their requirements. In Table 3.3, several of frameworks are compared for:

- Complexity of framework (X)
- Preciseness of requirements (P)
- Resource requirements (R)
- Focus on effectiveness (E1)
- Focus on efficiency (E2)
- Focus on consistency (C)
- Difficulty of implementation (D)

Table 3.3 Comparison of Performance Excellence and Quality System Models

Model	*X*	*P*	*R*	E_1	E_2	*C*	*D*	*L*	*M*
	L = good	[a]	L = good	H = good	H = good	H = good	L = good	H = good	H = good
Lean	L	L	M	M	H	H	M	H	H
ITIL	M	H	M	H	L	M	H	M	M
Baldrige	H	L	H	H	M	H	H	H	H
ISO 9000/1	H	M	H	M	L	M	H	M	M
Six Sigma	M	H	M	H	M	H	M	M	H
AS 9100	H	H	H	H	L	M	H	L	H
CObIT	M	M	H	M	L	M	M	M	M
CMMI	H	H	H	M	L	M	H	M	H
JCAHO	H	H	H	H	M	M	H	M	H

[a] Preciseness is situationally dependent whether high or low is good.

- Leverage with Organizational Learning models (L)
- Measurement of organizational performance (M)

Each of these Performance Excellence and Quality Systems models has both strengths and weaknesses across the range of attributes including how difficult they are to implement and the resources required for use. In some environments, certain models are mandatory, while in others they are selected because they have the best fit to their organization's culture and customer needs.

Each of these attributes is evaluated using a simple high (H), medium (M), and low (L). In almost all cases, the meaning of high or low is consistent for organizations, only the preciseness attribute is situational according to the type of organization and customers they serve.

To use this table for selection, you can construct a quick grading model where you simply apply scores of 1 for low, 2 for medium, and 3 for high (when high is good, invert scoring, i.e., low = 3), and then apply a weighting value to each attribute (as seen in the row immediately below the identifier). A quick calculation will show which model is the best fit for an organization's needs.

When you stand back and ask the fundamental question, "Why select Lean over other systems?" most simply, the answers are:

- It uses visual process improvement techniques and performance measures.
- It has a non-complicated approach and relies on simple tools and methods.
- It is not driven by numerical overload (over-analysis of data to the point of exhaustion—it is okay to move forward using subject matter expertise on simple opportunities).

A proposed LWL model is designed to integrate culture, values, and Lean into an overall model. Its structure might contain three basic strata: culture and values, core tenets, and operational systems, which include the supply chain, customers, and markets. The model is loosely fashioned after the TPS (Toyota Production System) model (Ohno, 1988).

3.5.1 Foundation of the Organization

At the bottom of the model are the culture and values of the organization. In this layer the roots of the organization must be examined and reinforced where positive examples are evident. Every organization has many examples of great behaviors that reinforce what is best in its culture and values. The fact that the world is rapidly moving or making a transformation needs to reinforce to the organization what it is doing best and identify what needs to be changed.

Figure 3.7 is an example of how culture and values are integrated and provide a bedrock for employee interactions, contributions, and behaviors. In the figure there are four example values at the bottom, each representing a distinct set of common beliefs of how all employees are supposed to act at all times. Sitting on top of the base is a set of behaviors that become the cornerstone for actions and key measures of organizational performance. These, when integrated together, form a common language, a common set of behaviors, and a singular vision for an organization's employee base.

Once the culture and behaviors have been defined and interlaced into the supporting organizational fabric, it is imperative to set simple, clear expectations for the organization's execution. The translation layer that sits on top of the culture and behaviors should be simplified into a very basic set of actionable statements called tenets. The next section will talk about what tenets look like and how you can use them, and offer a simple example of four tenets that cover the entire landscape of culture and behaviors.

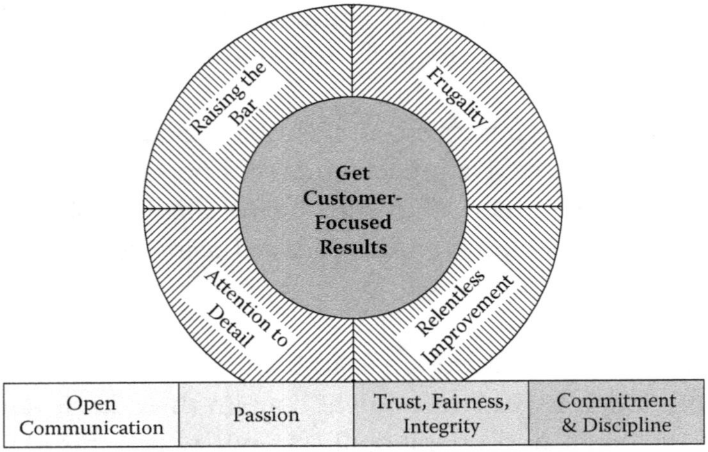

Figure 3.7 Integration of culture and values.

3.5.2 Core Tenets Align the Organization

Core tenets are positioned on top of the culture and values bedrock. Briefly, the definition of a tenet for this discussion is:

> A principle, belief, or doctrine generally held to be true; especially one held in common by members of an organization, movement, or profession.

The four example tenets align into simple, functional action statements that everybody can remember and use in their daily work. In Figure 3.8, there is an example of tenets that sit on top of and wrap together an organization's cultures and values:

- Be intolerant of bureaucracy, mediocrity, and waste.
- Make and meet your commitments.
- Execute flawlessly to world-class standards.
- Continuously improve in everything we do every day.

These four simple statements become the common relatable expectations that ground every employee in his or her daily work. The strength of the linkage of the four tenets to the elements in the end game is essential for relevance to the organization:

- Direction—Execute flawlessly to world-class standards.

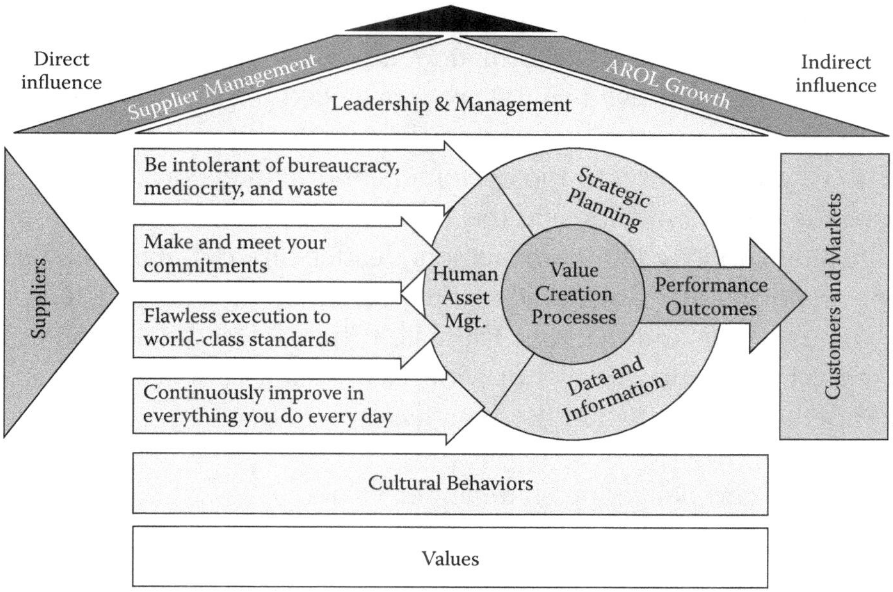

Figure 3.8 Lean Learning Organization Model.

- Course—Be intolerant of complexity, mediocrity, and waste.
- Speed—Continuously improve in everything we do every day.
- Relateability—Use simple language, be easy to remember, and be useable by everyone.

These become the short, one-floor elevator speeches anyone and everyone can use to communicate what is important as the organization races toward its end game.

3.5.3 Assemble the House

Thus far, there is an understanding and clear picture of the organization's culture and expected behaviors, and its daily expectations for performance, using the simple tenets that balance a learning model with performance excellence. In Figure 3.8, an example is offered that integrates core organizational functions and supply chain with customers and markets.

From the bottom up you can see the values and cultural behaviors form the bedrock of the organization. Above the bedrock is a layer of actionable tenets that set expectations for daily activities that focus on quality, speed, commitment, and waste reduction. These four tenets build the operational platform for the core organizational functions of human asset management, strategic planning, managing with data, and the entire value creation process. Their inputs (materials, information, and content requiring transformation into customer-perceived value) are introduced from the supply chain on the left side of the model. Transformed deliverables (products, services, or transactions) are delivered to the customers and markets on the right.

Above the entire model reside the leadership and management functions, which include supply chain management, leadership and management of the core organizational functions, and management of differentiated performance to customers and markets. Ultimately, transformed deliverables at a differentiated performance level provide measureable impacts for a sustainable competitive advantage in these areas:

- increasing and delivering shareholder value
- customer retention, growth, and loyalty
- growth in the organization's employee bench strength and retention
- robust and resilient supply chain optimization and sustainability

One of the strengths of this model is the seamless integration of the supply chain into the core value creation processes. Reflected in the model is the supply chain feeding into the organization's engine using the same operational tenets for performance. They act as an extension of the organization.

At the top, there is a leadership and management segment parallel to the organization's, but uniquely aligned over the supply chain's management structure. The essential point is that the organization's leadership and management team exerts expectations, disciplines, and performance outcomes on the supply chain exactly as they do on themselves.

3.6 Communicating to the Organization

> Leaders must wake people out of inertia. They must get people excited about something they've never seen before, something that does not yet exist.
>
> **Rosabeth Moss Kanter**

It is time to take your communication public. The previous chapters described what you would need to communicate: the content defining specific elements and the varying degrees of detail and vision that will be the right alchemy to achieve the results we are looking for. Now it is time to start to put this in motion, or as it is often referred to, it is "time to go public." By now, you have the specific elements of your deployment strategy laid out. You understand where you will focus the initial activities, the parts of the organization that you will need to engage, the early adopters that you are relying on, and a sense of the magnitude of the initial investments and activities you expect the organization to undertake.

With this as a backdrop it is important to set up Town Hall meetings to address as much of the organization as will be immediately affected by this new strategy. You should have socialized among your peers, your direct reports, and some early adopters the concepts that you now want to share with the extended audience. It is important at this point to realize this larger group of people will not have the luxury of this one-on-one time with you to understand the direction in which you are embarking.

As you construct and deliver Town Hall meetings, the references previously made to content become very relevant. You are trying to capture the minds and hearts of as many people in the audience as you can by reinforcing the need for change and by communicating your commitment to this path of investing in them to develop a better organization.

> People say that motivation doesn't last. Well, neither does bathing—that's why we recommend it daily.
>
> **Zig Ziglar**

The success of these meetings will not be determined by how much improvement is delivered in the near-term, but rather by the fact that you placed a stake in the ground that you are starting a journey. In delivering this message, you will need to be empathetic to the concerns, fears, and trepidations that the audience will have. You will address some of these with your personal resolve, the clarity of your direction, the urgency of the need to change, and the path that you have chosen to deliver a successful transformation of the organization.

Another criterion of success for these meetings is the chatter that is introduced to the organization through the Town Hall meetings. You will know if you have planted enough energy into the group to start the change if the chatter has sufficient volume and content to show that people understood the urgency for change and the direction you have outlined.

Once you have completed the Town Hall meeting, it is important to start campaigning in earnest. The word campaigning is being used in the political context and by following the model implying that you need to reach out to as many individuals as possible. This is advocated in a structured manner, so you can leverage combinations of various forms of meetings including:

- **Skip level meetings**—addressing the general concerns of the audience in a much more open and accessible manner. These meetings typically consist of individuals throughout the organization who hold non-managerial positions.
- **Brown bag meetings**—dealing with specific subject matter such as training content, project selection and definition, metrics, communication, and sharing of results. These meetings typically include people looking to get more insight into the next steps or clarity on the direction. The audience can be a wide range of functional representation and various levels of responsibility within the organization. For this reason, it is much better for the brown bag sessions to have a well-defined subject area advertised in advance to form the nucleus of the conversation to help keep things on track for the one-hour or half-hour meeting that you host.
- **Stand-up meetings**—an excellent way for various organizations to demonstrate initiatives that they have undertaken with their Lean projects to other groups. Here the audience may often be some of the early adopters who have been identified sharing successes that they have had or groups that are experimenting for the first time with the application of Lean technologies or practices into the work that they do.
- **Large group meetings and discussions**—the audience here typically represents site-specific congregations of employees who may or may not have been addressed during the original kickoff meeting or the original Town Hall sessions. There is a sense of caution with large group sessions because the content that you are delivering will vary little in the early phases. Repeating the message is important at this phase, but be careful not to do it too often en masse simply because it will prevent individuals from hearing anything different that allows them to grasp the concepts. Breaking it down into smaller sessions helps people address their specific concerns and allows you to understand better the level of resistance, fear, and trepidation associated with the changes you are proposing.

Recognizing that 80% of communication is nonverbal, the most effective means of communication for you comes from your demonstration of behaviors as they relate to LWL.

> The most important thing in communication is hearing what isn't said.
>
> **Peter Drucker**

How should this manifest itself in your daily activities? Since time is the most precious commodity you have, your use of time and application of it to the LWL initiative will be viewed by the organization as the degree of your commitment. Therefore, the need to weave the elements of LWL into your daily fabric is critical for effective communication. The tangible evidence of these elements in your daily activities will reinforce the message to the extended group at a rate unavailable to you in any one-on-one discussion.

Things to consider demonstrating in your daily activity are:

- evidence of your ability to identify waste
- the repertoire of anecdotal evidence of improvement
- incremental indicators of success that long-term improvement projects deliver
- near-term training and deployment statistics

Again, recognize the significance of porpoising in your conversations with people, oscillating from macro to micro and back, so that they can find the tangible elements that excite their particular concepts of how to deploy LWL.

The messages of this strategy have to be similar to a political campaign. The simple things that are reinforced repeatedly at various levels in the organization are critical to the success of LWL. Reinforce the message that this is an investment in the people doing the work, it is focused on improving the capabilities of the organization, and it is committed to delivering a better experience for your employees and your customers. These simple themes must be repeated.

In later sections, the concepts of management standard work and visual indicators will be integrated into your work. Those elements will reinforce the message, but for now people will simply want you to demonstrate in all aspects of your communication and engagements with them that you are committed to the transformation, its plans, and the tools and methods of LWL. By constantly repeating your understanding of the direction, the efforts necessary, and energies that are required to make this strategy a success, you as a leader will provide the necessary guidance to enable them to commit and start the change with you.

Key Messages in This Chapter

1. Don't overdesign the architecture for change—focus on intent not content…the direction is more important than the details.
2. The evolution that your organization will take is the product of thousands of small changes. You can't forecast the sequence or the cascading impact that each will have. There is no single approach, path or sequence for this journey.
3. Your visual image of waypoints along the journey will help others to see the potential.
4. If you have trouble setting a target remember that "twice the good" or "half the bad" are good starting points.
5. The amount of CCI you have will go a long way to determining the rate of improvement you are capable of implementing.
6. Your SME's and early adopters are a great way to determine your near-term launch points.
7. Base the transformation plan on change models that fit your organization's culture today—consider the three-layer model of Lewin, Kotter, and Hiatt.
8. There is a sweet spot in the deployment plan where it aligns with the organization and the first-tier or direct suppliers.
9. Choose the right performance excellence and quality system model space—the Lean house of quality is a great fit.
10. No matter what your plan looks like it will take twice as long to deliver half of the results. Remain focused, the journey is long but the rewards are a game changer!
11. The inertia in the organization will tax your efforts.
12. Not everyone will understand the message…you have to adjust your continent to the varying audiences that you need to reach.
13. You will quickly see pockets of success, leverage these anecdotal indicators of success across the largest possible segment of the organization.
14. You have an overall perspective of how your deployment plan will be executed including scope, near term deliverables, and next area of focus.

Challenge Actions

This section is about Planning the Transformation. It focuses on having a vision of the end game and creating a high-level plan to bring the gap between today's As-Is and tomorrow's Should-Be organizational environments. The actions address the identifying end game, refined through the VoC, and practical application of real-life constraints and conditions that bound all organizational change. Choose action items that capture what is important to your customers, critical to crafting a practical end state, and relevant to designing the transitions required.

Challenge Actions – Design the Transformation	*Points*
1. Spend 30 minutes doing personal investigation on an organization you know that embodies many of the traits you would like to see in your organization. Seek to find out what they do, how they made a change to get there, how long it took, and how much effort it took to get there.	30
2. Go to a whiteboard and draw a picture or table of must-have traits for your organization. Compare this to your mental picture of the end game—does it meet the necessary and sufficient test?	20
3. Construct an outline of what the future state could look like with, zero or minimal defects (in the data, product, process, or exchange of information). What will you do with the additional capacity? Does your business grow at the rate that will consume this extra capacity or will you be doing "more with less"?	25
4. Conduct a staff exercise where each member has 2 minutes (they will all exceed this to about 5 minutes, but institute the boundary anyway) to describe their view of an ideal state at a given time horizon (they set the horizon). Use the statements to converge on common themes and views of the ideal future state.	25
5. Ask your leadership team to evaluate the distribution of their employees according to the deployment model from innovators through change resisters.	15
6. As leader, evaluate your leadership team looking for where they fit in the deployment model and what is going to take to get them where they need to be.	20
7. Build a list of personal work processes upon which you can improve when you have discretionary time. During the course of the book, you will pull items from the list to do work.	20

8. Start to identify employees that fit in two categories: early adopters and natural Lean leaders.	15
9. Write honest answers to these questions: a) Are you capable of seeing OTF? b) Do you think it is important the organization sees OTF the way you do? c) Are you eager for a genuine cultural behavior within your organization?	15
10. Make a list of easy wins that will increase your CCI. These should be easily accomplished, provide effective demonstration of skills and impact on the organization, and can be replicated easily in other parts of the organization.	20
11. Make a list of the examples you will use to describe the potential of these changes to various audiences considering what it means to the individual contributor, managers, and organizational leaders. These examples will help you in delivering a clear message with each intended audience as well as add considerable CCI to you as they become reality. (The added benefit is that having the courage to describe some of these changes will help solidify the actions required to deliver them.)	20

Chapter 4

Preparing for the Race

> It's not the will to win—everyone has that, but the will to prepare to win that makes the difference.
>
> **Bear Bryant, 1913–1983**
> *Head Football Coach, University of Alabama*

In this chapter, deployment is starting to happen. It follows the Planning the Transformation Phase with deliverables including a deployment roadmap that uses the work defined in this chapter to deploy knowledge, skills, and expertise across your organization.

The quote from Bear Bryant is precisely the subject of this chapter. Preparation for the transformation is essential to long-term success and sustainability. The will to prepare is the key to transformation—the willingness to plan and invest in an organization's greatest assets, its employees.

This chapter will explore various levels of investment in training and skills acquisition, where they should occur, and the speed at which they should be made. Investments refer to training for skills acquisition, positioning the right people in the right place in the organization at the right time, and dedication of time to practice and use of the new skills.

Figure 4.1 guides you into the third phase of the transformation model—the Prepare phase. The output out of this phase will be newly trained or more skilled workforce, directional activities for process identification and focus, building Lean leaders, and knowledge experts identified and skilled in the ability to deploy Lean practices and organizational learning methods across the entire organization.

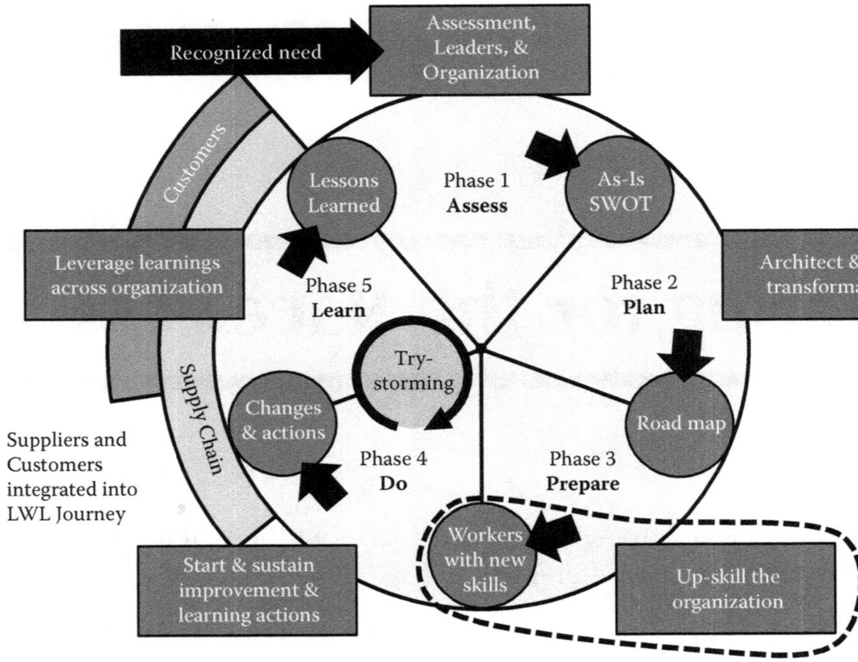

Figure 4.1 Prepare the organization, phase 3—transformation model.

Discussions will explore deployment reality, setting expectations, enrolling natural leaders, and communication across your organization.

Table 4.1 briefly explores the learning attributes of two adult learning styles. The learning styles—experiential and simulation—are used to highlight two of the more common approaches for educating working adults.

Experiential learning is based on a significant portion of the employee base who best learns by doing through trials, failures, and success of doing the work.

The following quote from Anne Sullivan summarizes it best—you cannot force learning on somebody and expect it to stick. They must be engaged with the execution of the work, get the sense of what works and what does not work, be able to fail successfully, and be able to learn and internalize the experience they gain.

> Don't be afraid to fail. Don't waste energy trying to cover up failure. Learn from your failures and go on to the next challenge. It's OK to fail. If you're not failing, you're not growing.
>
> **Anne Sullivan**

Simulation is the practice of artificially contrived environments with working practices that emulate the real process or task that someone is supposed to be learning. Employees are introduced into the simulation, provided artificial constraints, and allowed to fail without the potential to harm ongoing operations.

The premises for both of these learning styles is the same—learn by doing. The difference is experiential is faster on the learning because you are immersed in the actual process with all the real-life pressures and nuances, but failure could impact the organization. The simulation learning style requires students to make mental transpositions between a simulated environment and the real process. The speed of learning, or the acquisition of skills, is a little slower because they are not working on the actual process and they are missing all of the real-life consequences. However, failure is transparent to the organization because they can afford to try, fail, try again, fail, and repeat this as often as necessary to become proficient.

The balance of this chapter, which is all about preparing the organization, will cover the high-level setting of expectations for organizational learning, the reality of deployment (not everyone is in the same space, or will get to the same place at the same time), enrolling natural leaders to start the improvement fires, and resolve the selection of improvement systems and improvement tools that are used for the transformation.

4.1 Set Expectations

> Reward those who Do, Train those who Can't, Replace those who Won't.
>
> **Kevin Henn**

Here is the conundrum: it is so easy to see the waste in others before you see the waste in your own processes, but you know you should look inside first. We have all said, "If only they would improve X, Y, or Z, then we could improve our processes." These wastes are generally seen as process noise, visible waste of motion, input, energy, or just plain poor process execution that results in defects or poor service.

Managing your expectations and, as a result, the expectations of the organization is a critical part of this transformation. What is achievable and how quickly you can achieve it is the question everyone else in the organization is asking.

Let us put this into context. While dealing with a successful organization you realized that the ability to learn is not something that is naturally fostered or nurtured. Lean principles and concepts may or may not be well grounded within the fabric of the organization. So, while you may have years of success under your belt, you are starting from scratch with this endeavor. Just as a person going through physiotherapy realizes that the end goal may be a level of performance capabilities above what he or she has demonstrated historically, the first step in the journey will be painful and slow. In addition, so it is with this organization's rehabilitation to experience a slow journey.

Let us explore another physiological comparison. Doctors inform us that it takes 3 to 4 hours of exercise per week to maintain good cardio health. We have also come to believe that it takes as long to get back into shape as it took to get out of shape. Let us look at the average tenure of your employee. Assuming four hours of exercise per week, 50 weeks per year, you need to invest approximately 200 hours a year for each employee to keep diligent in his or her learning capability. While most organizations have deployed training through the course of the years for their employees, the amount of learning involved has been minimal. Continuing this thought process you can see that the average tenure of your employees is 10 years and you would have to invest 200 hours a year to keep current in their ability to learn, where approximately 2,000 person-hours are out of shape.

Another way to look at this is to explore some of your previous successes. In many of your previous endeavors, you only needed one group or one individual to accomplish the task and deliver the expected results. For LWL, we need the majority of the organization moving in the same direction to deliver the expected improvements. Therefore, it is easy to see that the level of training, education, and practice distributed throughout the entire group is unique to LWL.

- How many training sessions have been held?
- What percentage of the team is demonstrating competences with the tools?
- How many individual projects have been developed and completed with sustainable improvements?

These are the most basic indicators that you will use to determine the rate at which the team is developing expertise and capabilities in LWL. Remember, this will be confusing to many people because they will want to understand progress to the endpoint. The notion that you have to invest training to develop the competence within the organization to achieve the desired results is again a hurdle for many people. Your responsibility at this stage is to ensure that communication of the indicators of success is what the teams are measuring. The rate of adoption will help you determine what the limiting factors are for additional growth. Some of these may be as simple as hosting more training sessions or as complicated as hiring consultants to help with specific problem definition and Kaizen type solutions. Again, your role in this part of the deployment is to inject the correct amount of energy on the right dimension of the evolution.

4.1.1 Define Success Goals and Measures

> You can have data without information, but you cannot have information without data.
>
> **Daniel Keys Moran**

You are driving into defining and setting goals in key measures in this section, but it is important to recognize goals can be set up multiple levels of granularity according to audience and phase of transformation. The first discussion is about setting top-level goals, expectations, and key measures for the transformation as a whole, and setting specific expectations for workforce regarding up-skilling and deployment. This is the first critical, full organization engagement activity and it must be clear what is expected and how we are going to measure the rate of progress in change against those expectations. In later chapters, there will be discussion about setting detailed goals, expectations, and key measures again, but they will be at a granular level appropriate to improvement actions in targeted locations within the organization.

As you consider developing and implementing goals and measures, you are looking at evidence of proliferation of the ideas across the broadest sense of the population. You have to have the discipline to continue the investment in the individual because it is the individual capabilities that will determine the rate at which the organization evolves.

In the early stages of deployment, here are some simple metrics that can be used to help capture this:

- number of people trained on basic principles
- number of individual projects undertaken
- percentage of the population that has attended conceptual awareness training such as lunch and learn sessions or larger group meetings
- key tangible financial benefits achieved by using Lean techniques on specific problems

The idea of the metric that you are trying to measure may be quite different from the communication that you provide to staff. The extended audience at this stage of the deployment needs to start thinking of the order of magnitude of change available.

As the leader, you are looking for distribution across all levels of the organization. You are looking to see whether the small fires you have started are starting to spread. You are looking to see how much of your personal cash you need to invest in specific areas of the organization. This is considerably different from what you need the organization to gain from these early exposures to Lean projects. It is very easy for people to comprehend 5%, 10%, and 15% improvement challenges. It is very difficult for people to understand that 70%, 80%, and 90% improvements are available by eliminating waste. It is for this reason that we want small tangible evidence of these significant changes to help convince people that it can be applied to a much larger and broader scale.

Getting the team comfortable with challenging the status quo, attacking things that have always been done this way, and learning to embrace simplicity in an environment where complex solutions are quite often the norm are the types of psychological breakthroughs you are attempting to make at this stage of the game.

As you look to measure and reinforce these measurements, seek out examples of simplicity, seek out examples of improving that which is considered to be previously untouchable, and look for ways that people truly made their work experience better. With sufficient critical mass after a period of time, you are going to be able to start to see evidence of these improvements in the critical dimensions of the organization.

While most organizations that are successful already have mature measurement practices in place, it is frequently found that the measures are activity-based and rooted in measuring averages of performance. With the exceptions of highly repetitious manufacturing, service, or transactional processes, measuring variation and process capability is not as widely implemented.

One of the most critical barriers that prevent successful organizations from firmly establishing a sustainable competitive advantage hinges on effectively measuring and comparing performance. This does not mean good performing organizations do not measure well, but it begs some questions:

- Are they measuring the right thing?
- Are they measuring performance variation so they can tell how capable they are?
- Are they measuring the things that matter?
- Do the measures create tangible actions for improvement?
- Are the measures being reviewed and used as expected?

These, and a host more, just scratch the surface of determining how effectively they are measuring organizational performance. In this section, we are going to talk about the foundation (measuring outcomes), the heavy lifting (building and using the right indicators), and the review and use (taking actions).

Figure 4.2 builds a picture of a measurement system with three layers. Starting at the bottom is the foundation of performance measures. The next layer contains all of the measures used to gauge performance. The top layer is where the results are reviewed and analyzed, and actions are identified for future improvements.

4.1.1.1 Foundation Layer

Let us start with the first layer—the foundation of performance outcomes. This is the most basic element of this measurement hierarchy and it consists of only three outcomes: effectiveness, efficiency, and consistency.

Effectiveness is all about delivering a product, service, or deliverable that meets or exceeds the expectations of the customer. It is important to note that customers are used in the broadest sense here—customers can be the end customer (who writes the check), the next team or workstation in a process, a shareholder or stakeholder in the value chain, or the employees who are doing the work.

Efficiency is about achieving an effective deliverable in the most efficient way possible—think better, faster, cheaper. This is second in the outcomes

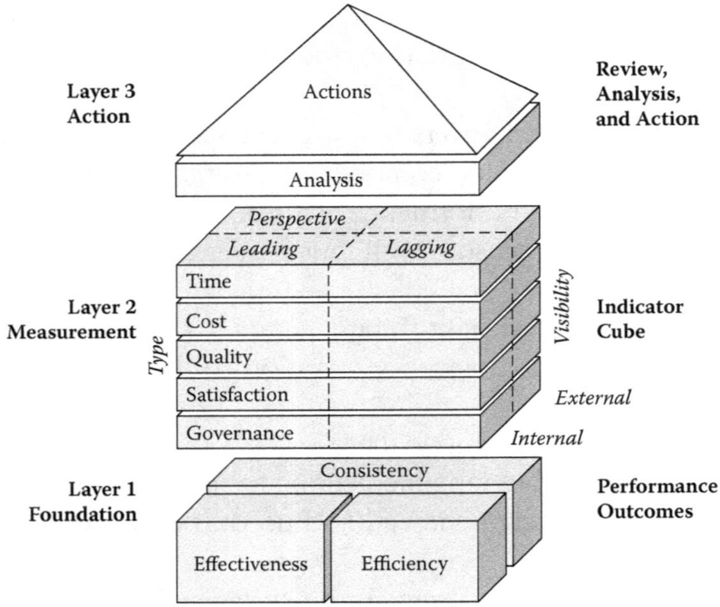

Figure 4.2 Performance measurement system.

because you do not want to be making more deliverables faster that do not meet customer expectations. Get it right first, and then do it cheaper and faster.

Consistency is an outcome and, while third in the order, it constitutes a vital element of how the operation is performing. Consistency is shorthand for process variability. This is third because first you have to figure out how to deliver the right thing to the customer, then you have to figure out how to do it at the lowest possible cost in the fastest possible time, and then you have to focus on being able to do that over and over and over the exact same way. The best part of consistency is that it reflects variation, both special and common cause, meaning that if you know how much and the type of variation that exists in your process you can determine the likelihood of being successful at delivering something to the customer that meets their needs. Once you understand that, and the cost of a failure to the customer, you can determine quickly how much to invest to improve the process outcomes. It is a simple economic model.

4.1.1.2 Measurement Layer

What sits on top of the foundation of performance outcomes? The second layer is the measurement layer of metrics used to evaluate and gauge performance. Let us call it the indicator cube because it has three dimensions:

- type of measures
- perspective of how measures are being used
- visibility of the measures related to the audience that will be seeing and using it

A brief exploration of each dimension follows.

Of the three dimensions, the type of measures is most complex. There are five typical types of measures:

- **Time**—Measures that have time as the unit being evaluated. Examples cover a huge range that might include process/production step timing, throughput times (TPT), project schedules, milestones, and so on.
- **Cost**—Measures that gauge performance in units of currency or in units that translate to currency such as a particular type of resource (e.g., person-hour). Examples include expense, investment, consumption, and conversion.
- **Quality**—Measures the conformance of a deliverable to a requirement or expectation. Frequently this is called goodness or fitness for use in many environments. Every process has execution or outputs, and each of these can be evaluated for performance to expectations. Naturally, examples are even more diverse than time or cost. Included in this type of measure are the "-ilities," such as reliability, repairability, repeatability, and maintainability (to mention just a few).
- **Satisfaction**—Includes measures for every entity contained in a value chain. This includes customers, suppliers, sub-tier suppliers, employees, regulators, stakeholders, shareholders, managers, leaders, and more. Frequently, these measures are subjective and require additional types of measures to remove the signals from the noise.
- **Governance**—Measures the compliance to legal, regulatory, or statutory requirements dictated to an organization because of the customer, market, region, or government it is required to satisfy.

These categories are the building blocks used to measure and monitor the performance of all core activities and deliverables. Almost all measures use one or more of these categories of measures. Using one of the categories alone provides information on levels, trends, variation, and comparisons, while integrating two of them enables measures of the rates of change or comparative levels (e.g., change in cost versus time, changing quality versus time, customer satisfaction versus quality).

The perspective of measure is best explained using a common context—leading or lagging.

- **Leading indicator**—A measure that enables a predictive view of performance in a subsequent activity or action. As an example, the outgoing quality level of a factory's production line is a leading indicator of potential field trouble calls; the lower the outgoing quality level, the more escapes into the field, the more failures at the customer point of use. This translates into more field service calls.
- **Lagging indicator**—A measure reflective of past performance and is used as a historic measure for analysis and trending in continuous process improvement. An example similar to the first is the first pass yield for deliverables through a series of process steps is the result of compounding failure levels at each task step, and it enables data segmentation to find the significant contributors to performance levels.

There are dichotomies in many areas of this transformation, and the measurement system is not immune. The fact exists that most measures have the ability to be both leading and lagging—the dichotomy of perspective or view for which you are using the indicator.

The third dimension to the measurement cube is visibility, and it refers to the audience that will be reviewing or using the measure. This falls into two large groups: internal and external.

- **Internal measures**—Used to communicate within that organization, are data rich, and should be designed to facilitate analysis and root cause corrective action.
- **External measures**—Designed to convey summary information that usually indicates a level and a trend. They frequently contain censored (or redacted) data or use aggregations of data to prevent inappropriate disclosure.

4.1.1.3 Action Layer

With the foundation layer of performance outcomes and the next measurement layer with the measurement cube that defines the dimensions of categories, perspective, and visibility, you get to the final action layer of an effective measurement system—review and use. It is where analysis, review, vetting of data, and decision making occur. Typical deliverables from this layer are racked, stacked, prioritized, and selected challenges and opportunities that require improvement actions.

The attributes of this layer include:

- cadence of review (timeliness and consistency)
- discipline of analysis (thoroughness and candor with data)
- decision making and action (data-led vetting and selection of actions)

Measuring performance is vital to improving and delivering value to the customer. Measuring the right things is even better. Measuring the right things and using the measures in a closed-loop feedback process elevates the effectiveness of the measurement system.

This book is all about Lean, and the premise of simplification and optimization has to be used on its own deployment. Do not waste time and resources on measuring things that do not matter. If the measure moves in the wrong direction,

- Do you notice it?
- Do you care?
- Do you react?
- Do you follow up?
- Are lessons gained through reaction, response, and corrective action?

If you cannot answer yes to each of these questions, there is likely an issue of either discipline or relevance. There are two more questions you have to ask yourself to lead you toward an answer—a necessary and sufficient test:

- Is this measure necessary to manage the process?
- If you continue the measurement (in concert with the other measures), will it be sufficient to manage the process?

If the answer to either of these questions is no, there is most likely a measurement system discipline opportunity. Too many organizations that are performing well waste extraordinary amounts of time and resources measuring things that just do not matter. They are not critical to customers, the organization, employees, or stakeholders.

In an honest necessary and sufficient test of performance monitors, the typical organization would likely identify approximately one-third of their monitors that are not responded to, and they offer little to no genuine insight into the organization's performance. To continue measuring them is a pure waste of time and resources, and they should be stopped right away. Be critical, be honest, and stop wasting valuable time and resources that do not yield benefit to managing the organization.

This book reviews techniques of comparing organizational systems and processes to thermodynamic systems, physiology, and the dynamics of physical science. Continuing the physiological analogies in the next subject, we draw parallels between biofeedback and organizational performance feedback systems.

Let us consider what biofeedback does. Based on a given level of performance, unique physical characteristics and measurements are used to focus on how the body can perform better and more efficiently. For organizational processes, the same concept holds true—there are unique performance measures that represent system and process performance and can be used to help improve performance and efficiencies.

Previously in Table 2.7, the parallels between physiologic measures and organizational measures could be seen and how they translate to physical performance and organizational performance.

You now have biofeedback versus EECO feedback, with appropriate parallels that apply across measurement systems. EECO feedback is grounded in this framework and enables a view across the organization to find points of support and points of disconnect and how we are managing organizational performance.

4.2 Deployment Reality

> You never change things by fighting the existing reality. To change something, build a new model that makes the existing model obsolete.
>
> **Richard Buckminster Fuller**

Let us talk about the reality of deployment. One of the hardest things to do in an organization is to change its cultural status quo, especially in a successful organization, from its current actions and behaviors to a new or modified set.

What is unique in this transformation process, because it is founded in the concepts of Lean, is that people are not being asked to radically change the way they analyze and critically review opportunities today. Most of the tools, methods, and techniques used in Lean are based on simple, fundamental engineering and logical practices. What is unique about the approach is the organization of the tools, how they are executed, the linkage between inputs and outputs of each, and the criticality of measurement systems.

Later in this section, enrollment of natural Lean leaders is discussed—their identification and selection, and the additional attention, resources, and training they will receive. They provide the critical link between the overall plan for transformation and the ability to make meaningful changes and increase efficiencies in the organization at the front line worker level.

The excitation energy required to get the Lean transformation moving will come from investing your personal CCI into change actions. The interesting thing about this CCI is that during deployment it grows in proportion with the increasing proliferation of Lean and its Kaizen events.

Where does this ever-increasing amount of CCI come from? Initially, it comes from the early adopters in the success that others achieve while deploying Lean practices. Your success in LWL is dependent on the adoption of these practices through the organization and the leverage of your CCI that comes from it. The dilemma will come in the early stages of deployment. The organization will see marked improvement as it capitalizes on the low-hanging fruit without fully using Lean tools and methods. Consequently, the amount of learning, in the beginning, will be limited.

Why is this a dilemma? The problem arises in that people have not necessarily learned how to approach problems differently but have made gains simply by accepting the premise that they are allowed to make changes. The dilemma in these results is that if underlying capabilities, tools, and methodologies to systemically approach problems have not been established there is a high risk that random improvement methods will be reinforced, and that achieving the results can be done using ad-hoc discipline with almost no organizational learning opportunities presented. You have to be very careful to leverage the success of these activities by focusing on the change achieved, not on the methodology used to achieve the result. While there may be much low-hanging fruit available, unless the organization learns how to apply these tools and practices systemically it will quickly lose momentum.

Another quandary that you will face is the notion that significant improvements in the performance characteristics of the organization may be perceived as success while not achieving a sustainable advantage. The capabilities and powers available by deploying the strategy of LWL provide significant progress in starting on some dimensions of your key performance parameters. This is where the use of a polar diagram will help you quickly gauge current and trending performance levels. By listing your critical processes along the axis of the spider diagram, you will note that your level of evolution will not be equivalent in all dimensions. Demonstrative success, or even leadership in one dimension, may relieve a lot of strain and stress from the team but it is important that you do not confuse success in one dimension as your overall goal. For example, significant improvement in OTD (On Time Delivery) while not addressing the issues of NPI cannot be conceived as success. Improved quality measured at points of manufacture or delivery versus the customer experience cannot be considered success. You cannot lose sight of the goal—achieving a sustainable competitive advantage by LWL.

In Figure 4.3, a simple representation of process performance and distribution for a range of core processes over time is shown. The importance

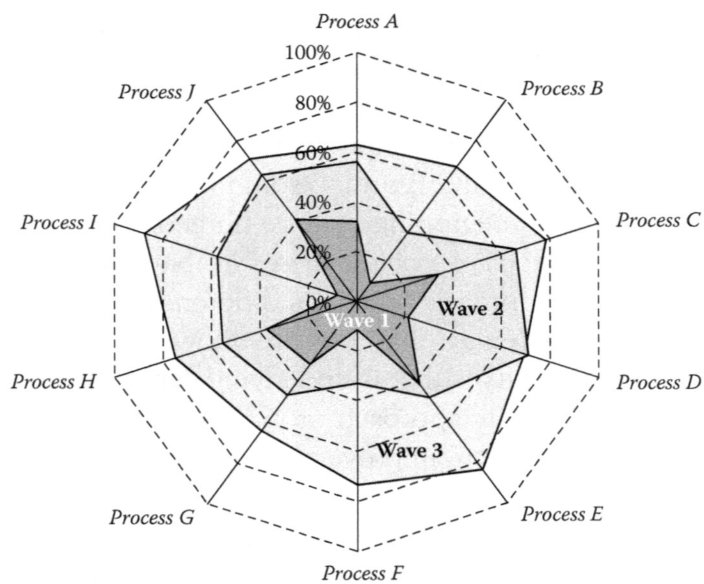

Figure 4.3 Distribution and evolution of core process performance.

of using a simple visual tool like the polar diagram is that you can rapidly show rates of progress or regress by core process as the organization matures. In the example, you can see Wave 1 (representing the first improvement cycle) as a small, highly varied set of results. There is significant variation at this point because the initial improvements covered a range from "low-hanging fruit" to "great wasteland" opportunities.

Wave 2 shows growth in all performance measures and demonstrates a more balanced set of results. Wave 3 bears two distinct messages: continued improvement in core process results and more balanced organization (when viewed against expectations). Sometimes, as seen in Wave 3 Process D, there is no change, even mild regression, in results, leading to the next layer of analysis—refinement and improvement. Additionally, you will see a wide variation around the performance results. It is also notable that the variation, while still wide in nature, is now being understood, measured, and evaluated for root cause. Early wins with recognition and expectations are being realized as Lean takes hold across the workforce. As noted in the curves, performance continues to rise and accelerate its rate of change, while the variation reduces as the curve rises.

4.3 Enroll Natural Lean Leaders

> A leader is one who knows the way, goes the way and shows the way.
>
> **John C. Maxwell**

John Maxwell summarized effectively the definition of a leader. This section talks about finding those employees who naturally lead their peers, and have the built-in desire to challenge and change the status quo. They believe that processes are inherently laden with waste, and they know that most employees are hampered by the construction or the constraints of the processes and systems in which they work.

As part of the deployment model and plan defined in Chapter 3, an additional look at the rollout roadmap is in order. In the plan, there will be a set of actions where certain employees will be identified who have particular skills and behaviors setting them apart from their peers. A group of working leaders, most likely to be acting as innovators and early adopters in the deployment curve (Figure 1.6), are the first group of targeted employees.

These employees are natural Lean leaders. They possess unique skills or characteristics that are best summarized as:

- Intolerant of status quo
- Seeking the easiest way
- Recognized by their peers as thought leaders
- Technically subject matter experts in their field
- Hungry to change their environment
- Willing to invest personal time for the gain of the organization
- Direct and constructively confrontational (related to process focus)

Depending on the design of the deployment model, you will find these Lean leaders have either direct, dedicated employees or the duties of being Lean leaders will be integrated into their daily roles. Either set of roles and responsibilities will suffice as long as the rewards, recognition, and acknowledgments coincide with duties and responsibilities.

A critical aspect for their success lies in two things:

- Clear definition of roles and responsibilities
- Clear definition of accountabilities and authorities

These natural leaders should get preferential training and special skills that enable them to be not only technically competent but also more effective in team chartering, problem identification and assignment, and program management discipline. Additionally, there should be activities that bring Lean leaders together that enable them to share best methods success and failure stories and their ability to become a cohesive team. Their inputs in process performance, process construction, and organizational effectiveness should be valued and integrated into deployment and rollout plans.

4.4 Invest in People

> Tell me and I forget, teach me and I may remember, involve me and I learn.
>
> **Benjamin Franklin**

Investing in up-skilling the workforce is one of best investments an organization can make to improve overall performance. Training, coaching, and mentoring to acquire Lean knowledge and skills is accomplished using traditional methods. Training will include different types of training for each level of employee, all the way from individual contributors to supervisors and managers to Lean leaders and ultimately organizational leaders.

Each level, or employee segment, is identified by his or her unique needs and potential contributions he or she will deliver. Examples of basic needs include:

- Frontline workers need simple tactic tools they can apply to their local tasks and processes.
- Supervisors and managers require tools that enable them to identify and lead projects, to build staff and guide teams, and for management standard work practices.
- Lean leaders are unique because they require additional technical skills, program management skills for the Kaizen teams, and communication skills required to communicate improvement results and findings.
- Senior leaders of the organization require training to be stakeholders, sponsors, and champions. They also require training on how to review, question, and approve projects as they move through their phases.

While training takes many forms, each organization will find and adopt training practices and principles for the highest return on investment unique to its environment and culture. There are multiple examples of simulations that enable higher training yields, student learnings, and embedding knowledge. It should be noted that training, while consistent across all parts of the organization, should be tailored to the kind of work that is principally done by employees. As an example, Lean training for design engineers would be vastly different from Lean training for transactional workers. While many of the tools are the same regardless of work type, how they are applied, typical outputs, and deployment practices vary.

Investing in people is the single biggest payback that an organization will achieve in this journey to building a sustainable competitive advantage. The investments must be timely, relevant to the organization, and exciting for the employees.

> Knowledge which is acquired under compulsion obtains no hold on the mind.
>
> **Plato**

In order to get the greatest return on training investment dollar, you need to focus on the highest rate of return for training spent. In Figure 4.4, you can see that the first round of training is extended to the specialized workforce called Lean leaders. These are the natural innovators and the explorers in the drive to eliminate waste. Following the Lean leaders are the supervisors and managers who provide first and second tier project initiation, improvement championing, and management review and approval. Training senior leadership and the executive branch is included to ensure improvement project alignment across the organization occurs and resources are invested at the right place and time.

Just behind the Lean leaders and management is the segment of mainstream employees. These are the members who touch process execution day in and day out, and as the simple tools of Lean get into their hands it affords the greatest potential return on training dollars.

Important in the training roadmap is inclusion of critical partners of the supply chain. In today's economic environment, many organizations use broad bases of suppliers, both domestic and international. Because they are important, integral members of the value chain, they are crucial players in the rollout of LWL. Installing new skills aligned with the wave of training, skills, and tools brings them closer to acting and executing like the host organization.

In Figure 4.4, training requirements are divided into areas including leadership and Lean leaders, mainstream adopters, and late adopters.

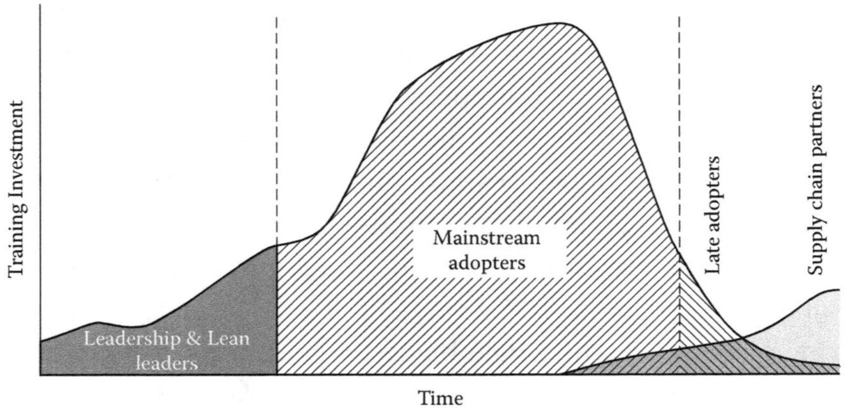

Figure 4.4 Roadmap of training investment.

Additionally, there is the supply chain segment that starts during the mainstream adopter group and continues onward. For each of these segments the needs for skills enhancement differ dramatically, and may use different types of training, or deployment, approaches uniquely tailored for them.

In the training model, there is a simple approach to build the instructor bench-strength, which uses a progressive acquisition of skills through four steps of increased knowledge and skills:

- get trained
- observe training
- co-teach
- teach

As deployment moves forward, it is important to find people who are willing to move up the knowledge chain. You will find the employee base is polarized with approximately 2% willing to invest themselves to acquire skills required to teach a subject. Do not over-expect everybody will move to the top, and even less will be willing to actively teach. Additionally, using an alternate Train-the-Trainer method of mentoring potential teachers is a critical approach to establishing a sustainable deployment methodology.

In Figure 4.4, a model relates organizational change and learning with a resultant change in organizational performance and reduction in variation. The change in skills corresponds to a lagging change in the momentum of the organization, or the shift in change. Trailing behind that you see the performance of the organization, while lagging in the beginning, quickly overtakes both the change in training and the change in momentum of the organization. At some point, it becomes an event horizon. This horizon denotes the transformation of the organization as it moves into a sustained competitive environment or advantage.

Training methods need to take into account the various styles of adult learning. Some styles require hands-on approaches, some best learn using academic class work, and some through engaged observation and analysis. While every organization's basic culture is different, the learning styles are similar in types and general proportions within the organization's employee base.

For example, an organization that designs, manufactures, and delivers products to customers typically has subcultures that include design or creative roles and tasks, disciplined execution behaviors in a manufacturing or operations division, and knowledge work and transactional behaviors in supporting organizations like sales, marketing, finance, and administration.

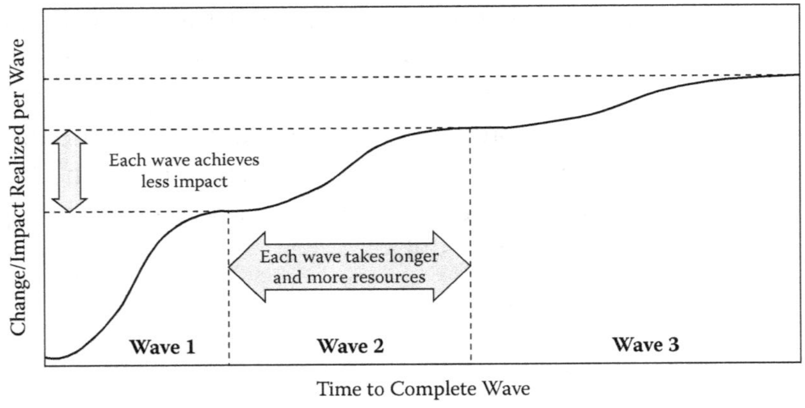

Figure 4.5 Change realized per improvement wave.

Training curriculums should not be one-size-fits-all, but adjusted according to the types of employee learning styles in the population. With adults, a typical training and deployment method is with simulation and experiential learning. These methods have some of the fastest and most effective returns for training investments (Good, 2003; Knerr et al., 2003).

Deployment occurs in successive waves across the organization. Figure 4.5 explores the concept of the waves of deployment and the amount of change that can be realized in each one of these waves or cycles. Remember, this is a journey, so there is no hard or fixed number of waves; the likelihood is that the waves continue occurring as long as the organization continues to demonstrate its ability to transform into a change-hungry organization capable of delivering a sustainable competitive advantage.

Impacting this figure are trends or outcomes from several other subjects in this book:

- deployment distribution model (Figure 1.6)
- training investment (Figure 4.4)
- categorizing waste opportunities (Figure 5.5)

As the organization completes each wave (there are only three waves noted in Figure 4.5) two things are immediately visible; the first is the duration of time to complete the wave gets longer with iterations, and the second is that with the amount of realized change and gains, each wave gets smaller.

Why are these two conditions occurring?

Wave duration gets longer because improvement actions are getting more difficult and complex to accomplish—the mainstream adopters and early adopters are already trained and in place. Now, work is being done on up-skilling laggards who are dragging their heels by nature. Many of the quick wins have already been identified and captured, and the improvement opportunities now require more cross-organizational work, implying the need for greater levels of communication, coordination, and effective program management.

The rate of change is decreasing because if you look at the tree of waste opportunity (Figure 5.5), you see low-hanging fruit that is significantly picked, and you are working in the great wasteland, which has a higher percentage of more difficult opportunities for improvement, increased need for data, and more complicated analysis tools that require investments of time and scarce resources. Things are just getting harder to find and to solve, but there is goodness in that story. The upside is that the majority of the obvious waste is gone, and now the search is on for the less visible, less obvious wastes. The downside is that people are investing more time and more energy in the improvement process.

In Chapter 3, an exit deliverable was a deployment architecture that included a training plan to deliver Lean tools and techniques across the workforce. Contained in those plans is the definition of which tools, concepts, and methods are expected to be installed and used by the organization as it moves forward and matures.

Central to these methods is an understanding of types of wastes. These are traditionally organized into seven categories including transportation, inventory, motion, waiting, over-production, over-processing, and defects. Each of these is briefly described in Table 4.1 and can be set into memory using the mnemonic TIMWOOD.

While the original seven categories capture and classify most traditional waste, recent discussions on additional categories of waste indicate that modern categories of waste may be missing from the initial seven.

In Table 4.2, the modern categories of skills, technology, and intellectual assets are summarized. They provide a simpler recognition of more complex waste categories.

A brief discussion on the modern waste categories follows:

- **Skills**—A category of waste that includes wasted human abilities or responsibilities given to workers who are not properly skilled or trained to perform the task. This particular type of waste is often overlooked,

Table 4.1 Seven Categories of Traditional Wastes

Transportation	Moving products that are not actually required to perform the process
Inventory	All components, work in process, and finished product not being processed
Motion	People or equipment moving or walking more than is required to perform the processing
Waiting	Waiting for the next production step
Over-production	Production ahead of demand
Over-processing	Poor tool or product design creating excess activity
Defects	Effort involved in inspecting for and fixing defects

Table 4.2 Three Categories of Modern Wastes

Skills	Human abilities are mismatched to the responsibilities given to workers—either over-skilled or under-skilled to perform the task.
Technology	Introduction of technology in processes that are not optimized or waste-free.
Intellectual Assets	Intellectual assets are developed and misused, under-used, or ignored.

but its potential to exist is especially likely when there are excess human resources in an organization being underutilized. This also is prevalent when an organization's tasks, responsibilities, and authorities are not effectively delegated in the organization. Included in this category of waste is the natural reaction of employees to under-perform because what is expected of them is far less than what they are capable of. It begins a tortuous circle of under-expectation and under-performance—a cycle that becomes fulfilling.

- **Technology**—It is the waste that is created through the introduction of technology in processes that are not optimized or waste free. The origin of this discussion is that it is easy to apply technology to a problem to make it more effective or make it more efficient. What is missing in the equation is the fact that technology, or automation, allows less disciplined focus in the removal of waste because the technology is so fast and so cheap that it is easier to automate a poor process than it is to streamline it and remove the waste first. There are examples in everyday life that we all have witnessed where automated processes are performing unnecessary actions or tasks because they can at a faster and cheaper rate than a human could. With such incredible technology power, it becomes easy to rely on its speed instead of removing unnecessary work, steps, or actions. In short, it allows lazy implementation of non-streamlined (waste-laden) systems and processes.

- **Intellectual assets**—This waste is characterized, much like the skills-based waste, by developed and misused, underused, or ignored intellectual assets. This may also be characterized by new concepts, designs, and solutions created by an organization, but unrelated to its current market and customer bases. Examples can be seen where organizations acquired another organization for the purpose of capturing its intellectual property and then burying it so competitors would not have the ability to see, copy, and use it against them. Additionally, in organizations that possess or exercise both original and applied research, the fruits of original research frequently are ahead of implementation possibilities and are shelved for potential future use, sale, or permanent elimination.

Seen in these 10 classifications of waste (seven traditional and three modern) are countless potential opportunities in most organizations where they apply and can be used to eliminate non-value added tasks. A maturing organization demonstrates the ability to characterize its core work processes and operating norms using these waste types so it can identify improvement tools, methods, and opportunities that will give the highest return on its investment.

4.5 Communicating to the Organization

> The strong man is the one who is able to intercept at will the communication between the senses and the mind.
>
> **Napoleon Bonaparte**

At this stage in the process, you are augmenting your ad-lib lines with data. You should be communicating up and down the chain in a language that resonates with the audience, using simple data examples, such as:

- percentage deployment
- people trained
- events held
- tangible impacts
- magnitudes of the changes

Examples of 80% improvement in a process step, examples of elimination of the opportunity for defect, and reductions in touch points in any step of the process will become the substance of what you use in your communication.

This is where your CCI will be deployed and where tangible evidence of early successes that you share with the organization will be the source from which it decides to adopt and commit. This is also the time when you can start to compare performance along the different critical processes that you supply.

Helping people understand that typical functionality, as defined by the organization, is not visible at the customer level or in the end-to-end experience is something in which they have to begin showing improvement. While they are now making progress at an individual level and on some projects, the idea of cross-functional contributions to more holistic solutions with broader engagements crossing traditional functional boundaries will be required to continue the evolution.

Key Messages in This Chapter

1. Preparation is the cornerstone of an organization's ability to make a transformation—it must have the right skills and tools available for the work.
2. Visible investment in the organization (tools and skills) is one of the requirements; another is the allocation of each individual's personal time to use and exercise the skills.
3. The transformation of the organization is directly linked to the visibility of the leader's commitment and daily exercise of the same behaviors expected within the organization—personally plan your investment in walking the talk.

4. Managing your expectations, the expectations of the organization, and the rest of the value chain is a crucial part of the transformation—focus on what is achievable, how quickly it can be achieved, and who is going to do the work.
5. During deployment, be selective in which engagements are chartered for improvement to maximize quick wins and key learnings.
6. Recognize there will be a wide range of performance in improved processes at the beginning of the deployment—it will appear to be uncoordinated, but it is only the various levels of implementation that create the range of performance.
7. Find and leverage natural Lean leaders in the organization; seek subject matter experts that are constantly upset with the status quo and have the passion to lead change.
8. Upskilling the workforce is one of the best investments an organization can do to improve its overall performance, improve employee satisfaction, and impact customer experience.
9. Training, while important across the entire organization, must be staged in its deployment to ensure the organization is not overwhelmed and incapable of using the new skills.
10. The seven traditional categories of waste have been expanded to include three modern categories of waste: the waste of workers' skills and abilities, the waste that technology introduces into process improvements through automation, and the waste of intellectual property misuse or non-use.
11. At this phase you have set near-term objectives that get the specific targeted groups moving forward.
12. Natural Lean leaders and early adopters have been identified and engaged in the transformation and how they will accelerate the deployment.
13. The organization is demonstrating that it is eager to invest in people (training, tools, and resources) who do the work and accelerate its transformation.

Challenge Actions

This section is about Preparing for the Race. It is all about taking the first step. There are key themes in this section about keeping it simple, investing in and training people to be competent in Lean tools and methods, and the fact that sometimes you have to kick-start an organization with a shove.

The actions in this list tend to be more activity based and aim at getting early, quick wins—those things that build credibility and put more credibility cash in Lean's bank. Choose action items that best fit the size and style of the organization, ones that will directly start Lean improvement actions and ones that begin (or reinforce) the art of learning through improvement.

Challenge Actions – Prepare for the Race	*Points*
1. Assess the rate of Lean tools deployment and training going on. Is it sufficient to deliver the results, or rate of change, that you are expecting? Build an appropriate action plan to address shortfalls, if any.	25
2. Review whether early wins (Lean process improvements) are visible within the organization, or do you need to integrate them into the communication plan to help propagate the message? Adjust the communication plan as required.	15
3. Draft a short list (5 to 8 items) where you see the pockets of resistance in the organization. They may be by function, site, or geography. Conduct a 5 Whys? exercise to determine what contributing factors are creating the resistance. Enroll the leadership where resistance exists and build a collective plan to address the root cause.	30
4. Perform an evaluation to gauge three things that are happening simultaneously: rate of training *versus* rate of learning *versus* rate of change. Answer one question: Is this happening on the timeline you expected or do you need to adjust your plan?	25
5. Record organizational changes or changes in management structure and personnel that are surfacing using the abbreviated approach of 3Ms (Man, Methods, and Materials).	20
6. Assess your measurement system's ability to accurately report on the organization's performance outcomes of effectiveness, efficiency, and consistency. If there are misses in measuring performance outcomes versus critical aspects of the organization's mission, develop a gap analysis, conduct a rack-and-stack analysis that prioritizes key gaps, and select the "vital few" to fix in the next 3 months.	25

7. Evaluate the measurement system's capability to invoke improvement actions — at all levels of workers, from customer facing and transactional workers to executive management. If the system is generally not auto-initiating, identify a core process where it does not work and conduct a system analysis to identify changes required to make the system flag issues that require correction.	20
8. Determine if "natural Lean leaders" exist within your organization (regardless of level) and are they provided special training and cultivation to maximize their impact within their work teams/groups. If this group of workers is unknown/unclear, develop and use a list of characteristics to identify the initial "vital few" and initiate a specialized training/coaching/mentoring approach that provides them additional skills. Set expectations for them to initiate improvement activities using the new tools.	25
9. Select a core business process and, using the 3 modern waste categories (Table 4.2), identify 5 opportunities for each type of waste contained in the "as-is" process, perform a rack-and-stack analysis to prioritize the opportunities, and select a 'vital few" to correct within 90 days.	20

Chapter 5

Go Improve Something—Start Doing

> The way to get started is to quit talking and begin doing.
>
> **Walt Disney, 1901–1966**
> *Inventor, Artist, and Organization Leader*

You have arrived at the step moving from getting ready to making a difference. Walt Disney says it best, "…quit talking and begin doing." You have moved from preparing and getting everybody trained and skilled. Now, you start applying the new skills to the real-life challenges and performance opportunities in your organization.

The output of this phase is changed and improved processes or systems. The task is the application of skills, the implementation of change, the elimination of waste, improvements in efficiency and effectiveness, and designs that ensure consistent performance over time.

In Figure 5.1, the transformation cycle indicates you are in the fourth phase, dedicated to starting and sustaining improvements leading to continuous learning cycles. This is where the rubber meets the road. Employees are trained, new skills are present in the workforce, core processes are identified, and the organization recognizes the various levels of opportunities from low-hanging fruit all the way up to the top of the tree. The outcomes and deliverables from this phase are changed and improved processes with streamlined, optimized execution.

There are several cautions as you move into Phase 4 of the transformation cycle:

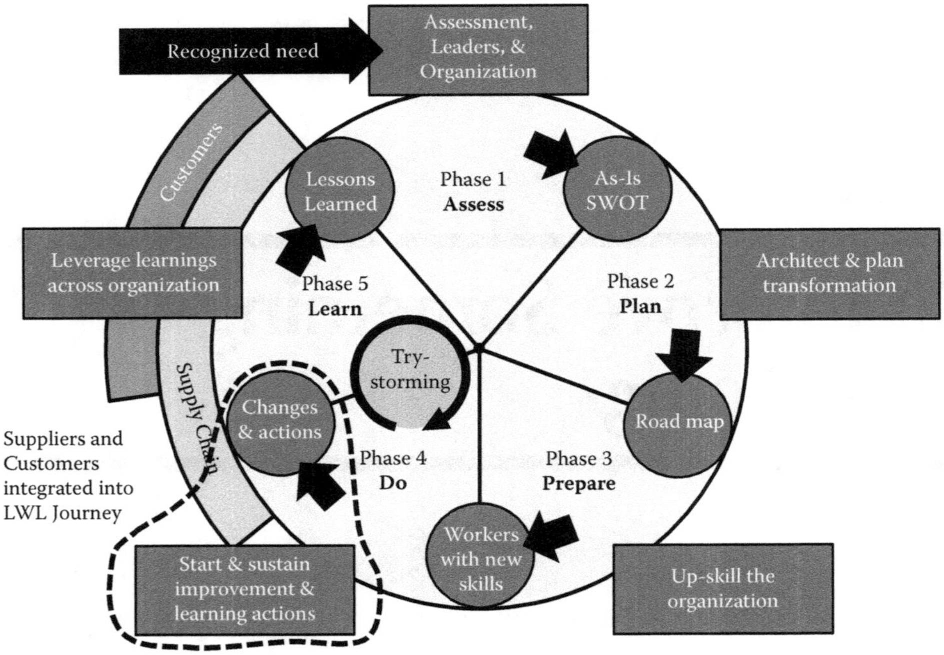

Figure 5.1 Doing, Phase 4—transformation model.

- There will be employees at all different levels of knowledge, buy-in, and energy level.
- Some members of the organization, while appearing to be engaged in moving forward, are really just treading water, hoping that this is a fad that will soon disappear,
- Some members of the management and leadership team will not fully buy in, and will give only superficial support.
- Many workers will be overburdened with fear—they have heard stories of job loss, replacement, or relocation as wastes are found and removed, leading to fewer people doing the same amount of work.
- Expect that some people will not be transparent and forthcoming in what they find or see because human nature says they do not want to be embarrassed because they have not worked using the best, or easiest, ways.
- Employees' resistance to change will be most visible because you are actually asking them to change the way they work, act, and behave.
- Stop using sparse SMEs on each Kaizen team; use the data with independent, effective analysis, and then vet the findings and conclusions with SMEs to ensure the team is getting it right. This is a much wiser use of their time, knowledge, and scarce resources.

Deliverables in this chapter will be results from actually doing work—improving things—by taking a newly trained workforce and having them apply their skills to new opportunities for waste removal.

5.1 Go Ahead—Give It a Push

> Take time to deliberate, but when the time for action has arrived, stop thinking and go in.
>
> **Napoleon Bonaparte**

Napoleon's quote strikes home with this phase of the transformation journey. You have taken time to understand your strengths and challenges, create your plans, and train your people. The time has arrived to take action and do something with what you have learned.

A push is often required to start moving a successful organization that is locked into its ways. When pressure is applied, expect that resistance will be present and that if you let off the pressure, the organization will revert to its original ways.

So, where do you start? There are two things to think about at this point: opportunities for quick process wins and opportunities that are low-hanging fruit. While they sound the same, they are different. In the first case, a quick win is an opportunity that allows the employees to practice their new skills and knowledge to solve problems—this is how they learn. In the second case, there are opportunities that are ripe for solutions and are generally the ones that have the greatest impact for effective execution. These are the daily problems everyone sees and they are frequently easy to find and fix.

Always stay focused on what is important to the organization, but remember in the beginning improvements will be small effort by individuals or small teams, and may be hard to translate into measurable impacts. That is okay—what is important is to get the ball rolling, to get the people moving forward, and to get the new tools in use so that others see their benefit.

What should you focus on? This question has several sides; each one is valid and has contributions to a successful organization. As leader, you will need to be able to shift the focus in the organization according to the capabilities, energy, and maturity of your team. In the following three simple questions you can see the dichotomies in what to focus on:

- Performance improvements versus learning?
- Practical versus theoretical problem solving?
- What will drive your personal CCI higher?

Let us briefly explore each of these questions.

Performance improvements versus learning? Centers on improving versus learning. While both are important in the beginning, quick wins are necessary to demonstrate value and impact. Learnings, while important, are a close second to being able to make a positive impact on performance.

Practical versus theoretical problem solving? Checks the balance of theory versus practical solution. Understanding theory, how things work, and how they are linked together is vital to successful implementation, but practical solutions are what people remember. Most people learn by doing. Practical problem solving is one of the strongest learning tools in an organization's toolkit.

What will drive your personal CCI higher? Focuses on increasing your ability to enable, leverage, and implement changes in the organization. Some of the time, the focus is about increasing your CCI so you can leverage your value to the organization for higher performance goals or help navigate the organization through a rocky path.

What about today? What should you focus on now? The most obvious answer is the thing that aggravates you, or the organization, the most. Logic says it is probably the lowest value-add task. Solve the nagging problems first and get them out of your way so the view into organizational performance has greater clarity and data drives improvement opportunity selections versus intuition.

Each of the five phases use prescribed Lean tools that are simple and effective. By design, each of the tools promotes organizational learning using team experiences, vetting of data and its results, and the wide use of team knowledge and background as SMEs to arrive at an effective and efficient solution. These are well described in many books by numerous authors; a few include *Lean Transformation: How to Change Your Business into a Lean Enterprise* (Henderson & Larco, 1999) and *Seeing the Whole Value Stream* (Jones & Womack, 2011). In these works, each of the basic Lean tools is broken down into how it supports and embraces organizational learning. The link is evident, and leveraging the tools for organizational learning will

Table 5.1 Dichotomy of Kaizen Event versus Continuous Process Improvement Project

	Kaizen Event	*Continuous Improvement*
Improvement focus	Small, localized improvement action on process or product-specific opportunity	Generally implemented at process and system level with cycles of improvement designed into methodology
Impact on organization	Small to moderate impacts to organizational performance realized through localized improvement efforts, in some cases significant impact to organization for improvement actions on core organization processes	Generally medium to significant impact to the organization, tied to implementation on larger or more complex core processes
Investment required	Very low to moderate investment, generally small team, short timeframe, low capital investment	Can be anywhere from low to high investment required, based on complexity of opportunity, speed of solution (urgency), or diverse nature and geography
Magnitude and duration	Short, fast, swift; for small, localized opportunities, only a few hours; for more significant processed investments, it can be small teams for a few days or a couple of weeks	Can range from short (weeks) to perpetual, depending on opportunity methods and ability to invest appropriately to bring closure to improvement actions

become commonplace the more they are used and exercised. Expanding on the tools link, Table 5.1 explores the differences between Kaizen events and continuous improvement projects.

Let us focus on one of the simple tools in the Lean toolbox that enables or supports organizational learning. Kaizen events are designed to be short, rapid improvement actions that focus on a well-scoped and bounded opportunity or challenge. The basic structure of the Kaizen event involves the phases of discovery or investigation, data analysis and root cause identification, rapid solutioning (try storming), and implementation of measurement and control systems.

As the leader of the organization, there are important behaviors to demonstrate as you guide the transformation. These include:

- using Kaizen eyes on everything you see,
- insisting on OTF for all work,
- rigorously adopting the concept of asking "Why?" five times to be thorough,
- pushing for true root cause corrective action, and
- ensuring error-proof solutions are standard practice.

Kaizen eyes and OTF have been discussed in other sections of this book. Additionally, asking "Why?" five times and searching for true root cause have been explored. Now we are going to introduce the practice of continuously striving for implementation of error-proof solutions.

> A common mistake that people make when trying to design something foolproof is to underestimate the ingenuity of complete fools.
>
> **Douglas Adams**

The concept of error proofing, called Poke Yoke, is a core philosophy and practice that is expected in every design whether the design is a product, process, service, or transaction.

Error proofing is defined as:

> The practice of designing a deliverable or its value creation processes so that it cannot be made or executed in such a way that allows an error or escape to occur.

A few everyday examples of error proofing include:

- The difference in the size of the nozzles between diesel and gasoline fuels prevents incorrect filling of diesel in a gasoline car.
- An over-temperature shutdown switch in a clothes dryer shuts the dryer off if the dryer becomes too hot for any reason.
- Ignition keys for vehicles that are double cut on both sides ensure that it does not matter which orientation is used by the operator to insert the key and start the car.
- Polarized electrical sockets in the wall prevent potentially ungrounded insertion of an appliance leading to a safety hazard.

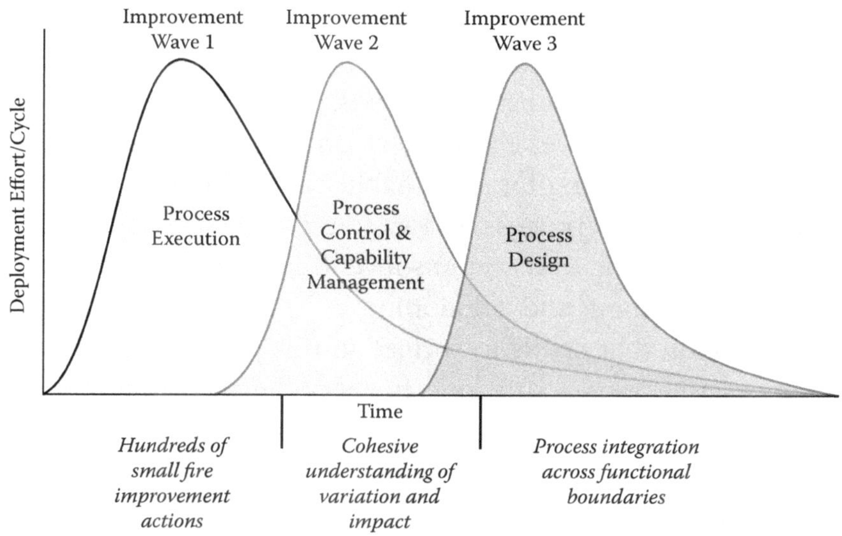

Figure 5.2 Maturing focus for each improvement wave.

This approach used during the design of the deliverable or process is essential to increasing quality and efficiencies, and decreasing costs. It is a behavior that, as leader, you should be pushing every time an improvement solution is presented. By repetitively asking how to build something so it cannot fail, eventually it will become embedded into the nature of the organization. You should set it as one of the basic calling cards for successful review and approval of solutions and their implementation.

> We should work on the process, not the outcome of the processes.
>
> **W. Edwards Deming**

In Figure 5.2, a successive set of improvement waves implies deployment maturity across the organization as the transformation roadmap is executed.

The first wave focuses on process execution where opportunities, characterized as low-hanging fruit and the great wasteland, are best found and waste eliminated. Organizationally, this is best characterized as thousands of small improvement actions, but with low levels of alignment and coordination across the organization.

The second improvement wave focuses on controlling processes and system, and monitoring their capability management. The design of this wave is represented as a successive level of improvement, use of more complex tools for analysis of opportunities at the "tough ground" level, and in the construction of processes. The organization is becoming more focused on assuring cost-effective solutions that are consistently capable of delivering results that meet, or exceed, customer needs.

The third wave of improvement focuses on process design where the concept of error proofing for sustainable competitive advantage surfaces quickly. Evidence of process and system integration across functions becomes readily apparent, and systems-level thinking becomes the norm of process analysis, planning, and solutioning.

There is a caution that needs to surface in this discussion. Senior leadership in most organizations is driven to focus on large, complex issues and consequently they want to find large opportunities to solutions. The distraction is that large, complex opportunities drive complex solutions. The approach advocated in this book is one of starting with quick wins that build confidence and proficiency. Senior leaders must take note that if they push too fast or too hard in the beginning, the organization is likely to trip and stumble, potentially leading to a stalling of forward progress. They need to save the focus and energy for big projects until at least wave two, but most likely until wave three.

5.2 So You Think You Know How It Works

> Don't look with your eyes, look with your feet. Don't think with your head, think with your hands.
>
> **Taiichi Ohno**

You are the leader—you manage the organization and its operational performance and you think you know how it works. There is an old axiom that goes something like

> The higher you are in the organization, the farther you are from the process, the less you know how it works.

Regardless of the organization, regardless of the technical capabilities of the leaders, the way a process is designed and implemented is always different from the way it actually is executed. Just witness work being done to see where waste resides. Gemba walks provide the familiarity you need to be able to effectively charter, staff, and manage an improvement team.

If you want to know how something works, you need to take a walk on the floor. You must see the flow and actually walk the process. You need to go where the process is done, where it is executed, where you can talk to the employees who live with the process day in and day out. Something cannot be improved based on presumed knowledge or the way it is supposed to work.

Gemba walks are one of the first critical activities in a Lean improvement effort. It is the opportunity to see the process as is. Gemba walks are your first line of understanding. They are an effective tool in being able to see the flow, to watch where the work is being done, and to understand the tasks in the value transformation of the deliverable.

Gemba walks are not casual walks. They require preparation, they are methodical, and they take dedicated time and attention to be done correctly. It is critical to make sure to capture the as-is state in a well-documented and accurate record. After you have done your first one, you will be amazed and impressed regarding what you observe and what you learn. Today, you believe you know how it works—wait until you get on the floor.

There are basic rules you need to demonstrate when you conduct a Gemba walk:

1. Treat workers with respect. They are working their best in a process that frequently they do not control.
2. Respect the process; it was developed, designed, and implemented with the best of intentions.
3. Walk the flow with Kaizen eyes. See every step, ask "why" questions, ask "how" questions.
4. Record what you see exactly, and ask questions of clarity not challenge.
5. Be prepared, know what the categories of waste are, know what waste looks like, and know how to find it.

Using these few simple rules to manage a Gemba walk, they begin to allay the natural fears when a manager or senior leader comes down to the floor and observes a process in its execution. By respecting both the workers and the process, it shows your earnestness to be open, candid, and inquisitive. It demonstrates focus on facts, objectivity of observation, and passion for trying to understand before you try to improve. One of Stephen

Covey's concepts from his book, *Principle-Centered Leadership*, holds especially true here:

> Seek first to understand before being understood. (Covey, 1991)

Here are some observations and hints when doing Gemba walks:

- They are exciting because you will see the process in all its glory, exactly as it is actually done.
- Wastes and opportunities will jump out at you. Do not overreact or condemn—seek to clarify and understand.
- Dress and act just like those who are running the process. Try to remove the artificial barrier of management that inserts itself in process improvement.
- Record what you see exactly; listen to what they say happens as exceptions occur and how they work around the process.
- Do not offer solutions yet even though they may be apparent because you need to have process owners enrolled in how to fix it.

At this point, let us reinforce a simple tool for managers to begin breaking down the opportunities found during the Gemba walk. The simplest for a leader to create and use is called a Fishbone chart (some call it an Ishikawa diagram). It is built simply with a backbone and ribs that look like the bones of a fish. While there is a large range of potential classifications for the ribs, these five ribs cover the majority of potential causes for the opportunity at the head of the fish. They are:

- Man—causes that result from human interaction with the process.
- Methods—prescribed or work-around methods that do not deliver effective, efficient, or consistent process outcomes.
- Machines—tools used (physical, mental, or virtual) to execute processes.
- Materials—raw goods (physical, data/information, services, or others) used in the value transformation processes.
- Management—oversight, controls, authorities, and approvals that ensure processes receive guidance, resources, roadblock removal, and control.

An even faster alternate form uses 3Ms: man, machines, and methods, on the "back of the napkin."

In either case, using a Fishbone diagram is a simple and very easy to use tool that enables logical analysis and breakdown of potential root causes. Enlisting subject matter experts to help build and vet the fish is a wise use of their sparse time and deep knowledge.

5.2.1 Entropy and Noisy Systems

The 1990s was a time of challenge to the automotive world. Significant competition emerged between Pacific-rim and U.S. manufacturers. An executive in one of the Pacific-rim manufacturers was quoted as saying (unattributed):

> Western companies use great people in poor processes and get average results; our company uses average people in great processes and gets superior results

The messages in this short statement reveal legions of information about organizational leadership for effectiveness, efficiency, and consistency. The context of getting mediocre results from exceptional workers constrained by average processes is a tattletale on where they choose to focus investment of time and resources.

Simply put, Lean's focus is on finding and eliminating waste. The definition of entropy is energy contained in the system that is unavailable for work—waste. This quickly leads to the proposition that Lean is the antithesis of entropy.

Using the first law of thermodynamics as a proxy for any organizational system, you can extract that there is latent energy available in all processes. Lean focus is directly opposite in purpose and, when executed, seeks to find where latent energy is, determine how to capture it, and then direct the energy for additional value-added work. While this seems simple in description, there is a whole field of science that helps find and promote easy, simplistic ways of achieving this objective.

> It's only the last turn of a bolt that tightens it—the rest is just movement.
>
> **Shigeo Shingo**

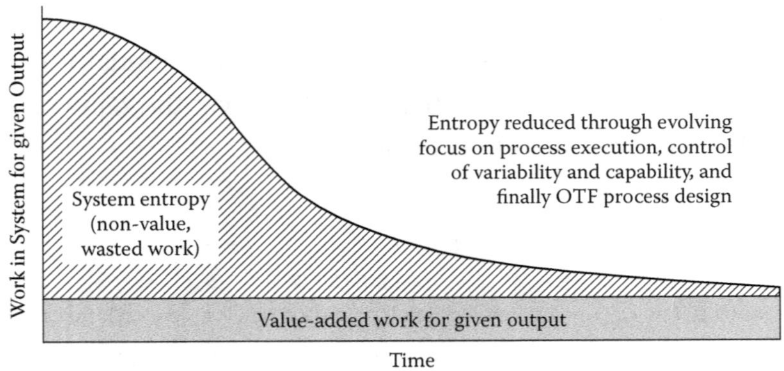

Figure 5.3 Work entropy in a thermo-economic system.

Figure 5.3 is a good example of how a work system acquires non-value added work, which is generally lost as heat in the system (in thermodynamic systems, heat is the proxy for waste).

The classic definition of entropy in a thermodynamic system is defined as:

> The measure of a system's (closed system) thermal energy per unit temperature that is unavailable for doing useful work.

A parallel definition of entropy in a thermo-economic system is:

> The measure of a system's (closed system) work capability per unit of resource that is unavailable for doing useful work.

What is the linkage between OL, the use of Lean, and the creation of a sustainable competitive advantage?

Throughout this book, improvement topics have been dealt with independently. Now it is time to begin overlaying each discussion and weaving them together into the fabric of the organization. Three targeted topics for integration include existence of waste in all processes, focused use of Lean tools and methods to identify and eradicate waste wherever possible, and organizational learning as a natural result from waste elimination activities. These significant points for integration are explored next.

First, waste exists in every system or process, no matter how small or how large it is. Typical wastes can be categorized in the seven traditional categories plus the three new categories of skills, technology, and intellectual assets. With almost 80% of all work done having no value (non-value add activities) for the customers it would be a perfect world if you were able to eliminate all of that waste, but such a utopia does not exist. Your tasks as leader and leadership teams are to help align, coordinate, and direct where the greatest value to the organization results from improvement investment.

Second, with the premise that Lean methods and objectives identify and remove waste, it can be seen that the linkage between entropy elimination and implementation of Lean methods is strong. Unique to each organization's deployment is the characterization of the significant types of waste their processes and methods systems contain and then tailoring the Lean tools that best fit the waste removal. The implication is that Lean champions and Lean leaders must be identifying and coordinating of significant waste removal. Management's charter is to ensure that enough trained resources, proper tools, and allocated time are dedicated to make waste removal actually occur. Additionally, management and leadership are accountable for ensuring that improvements, once implemented, are sustained through appropriate policies, follow-up, measurement, and control systems.

Third, it has been shown that the use of Lean tools directly enables OL through their construction and use. Connecting the dots, from understanding system entropy through proper Lean tools implementation and the allocation and investment of improvement resources, a strong landscape for continuous improvement in an organization's culture and operating fabric emerges.

5.2.1.1 Noisy Systems

What is a noisy system? A quick definition for this book's purpose is:

> A system that is designed to allow excess variability (common and special cause) and distortion into its execution, thereby allowing faulty signals to corrupt its effectiveness, efficiency, and consistency.

Let us look at Figure 5.4, which provides a planning system transformation that occurred in a high-tech electronics manufacturing organization. The situation involves an organization and its supply chain partner. In the top half of the example you see multiple divisions within the organization routinely sending demand plans to the supply chain partner individually and uncoordinated. The supply chain partner reacts and responds to each incoming signal independently. The design of the system when viewed at the organization unit level is effective, but when viewed from the supply chain partner's perspective it enables large degrees of variation and

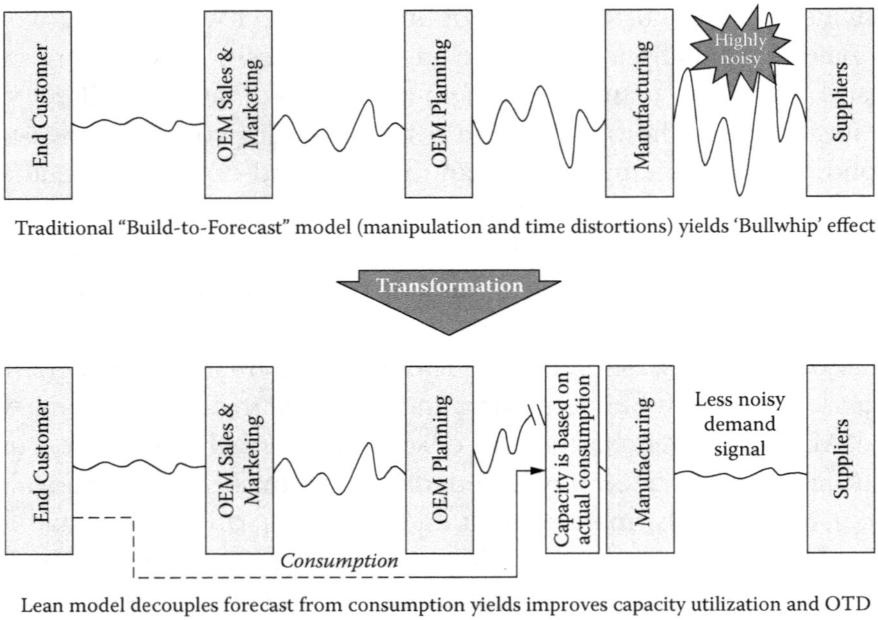

Figure 5.4 Noisy system, planning systems example.

conflicting messages—signals—to drive erratic demand plans. In its original design the system only took into account the originating organization, without regard to the impact of the supply chain partner. This erratic output is largely due to:

- Control input works to stabilize the system, and the noise input works to destabilize the system.
- Two inputs are independent.
- Two inputs may act on the system in completely different ways.

In the top part of the figure, you see very noisy signals (orders demand planning forecasting and so forth) that come from the parent organization but from different divisions or units within that organization. Each one of these divisions or units supplies its own demand, its own needs, into the supply chain. The supply chain responds according to the individual inputs but is confounded because the inputs are not prioritized, integrated into larger orders for economies of scale, or staged according to the needs of the parent organization. The extra effort and energy required by the suppliers manifest in longer lead times, product delivery errors, higher costs, and general disruption.

In the lower half of the figure, you see where the parent organization has instituted a new process of alignment where the multiple divisions integrate their demand into a single output and that output goes into the suppliers in a much more cohesive and unified fashion. The benefits of this are faster supplier response, lower costs, more consistent delivery to the organization's needs, and reduced internal efforts in the supply chain. As a reminder, the suppliers and supply chain are by definition part of the value chain. As such, they deserve to be integrated with core processes with the intent to simplify, streamline, and increase effectiveness and efficiency.

5.2.2 Automation Creates Techno-Waste

> The most dangerous kind of waste is the waste we do not recognize.
>
> **Shigeo Shingo**

What is techno-waste? For the purpose of this book, the definition of techno-waste is:

> Techno-waste is the small amounts of micro-sized waste that is contained in a process that has not been simplified or optimized, yet it has been institutionalized into a technology (software or hardware) and has only hidden the waste.

It is the accumulation of small bits and bytes of wasteful process steps (process "dust") into an automated process by using the sheer power of cheap technology. All processes must be simplified first. The best initial approach, and history shows this to be true, is to execute a defined process only with pencil and paper. The moment you use simple fundamental tools, you will find out where inefficiency is built into the design. It will trigger removal before you attempt to automate.

A strong parallel is in the designing of software or firmware deliverables. Because of the availability of high processing power and large memory capacity at very low cost, it is easy for programmers to develop additional steps (e.g., certain types of internal error checking routines, unnecessary redo loops, extraneous steps and counters, boundary condition bypass and loops, etc.) in code that would be unnecessary if developed with a minimalist perspective for design. They can afford to do this because machines are so much faster than humans are, and the time required to perform some of the steps happens in just a few processor clock cycles. This all points to a drift away from efficient programming discipline. The side issues occurring often can be seen as unmanageable, unmaintainable code, and slow error correction cycle (bug fixes) due to discovery effort required to unravel poorly written code.

> The first rule of any technology used in an organization is that automation applied to an efficient operation will magnify the efficiency. The second is that automation applied to an inefficient operation will magnify the inefficiency.
>
> **Bill Gates**

Technology creates entropy in processes and systems because cheaper technology can be used to automate a wasteful process quickly. Following this approach drives waste below the table, out of view, and creates the culture of allowable techno-waste to be designed into future solutions.

5.3 Data Will Set You Free

> As a general rule, the most successful man in life is the man who has the best information.
>
> **Benjamin Disraeli**

The cornerstone of any effective organization in managing and monitoring its performance is how well it uses data analysis and extraction of information. In this section, the data required, which stage of improvement, and what skill level is required to use the data for certain types of improvements will be explored. Avoid artificial savings measurements; recognize and reward positive changes in capacity and defect density. Time your measures development and introduction. Do not measure too early; do not jump on the "if it doesn't jingle it doesn't count" mantra too early.

Figure 5.5 refers to the complexity of the data required for the kind of waste opportunity present in an organization. Using a simple segmentation approach, there are three easy to recognize segments. Each of these zones of opportunity is briefly described, including typical analysis in simplification tools.

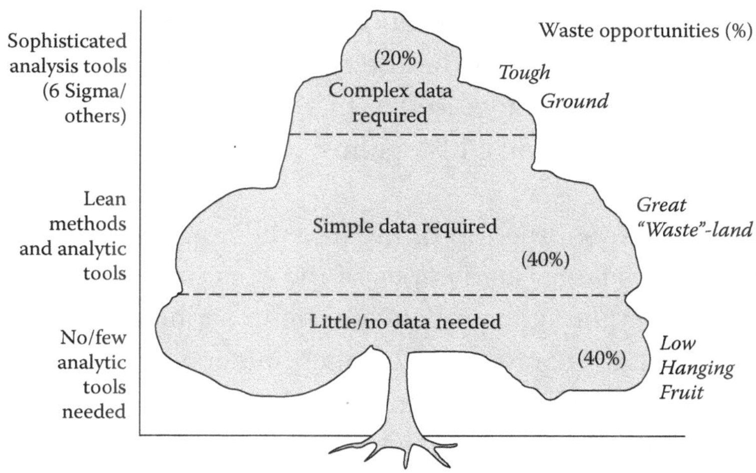

Figure 5.5 Tree of waste opportunities.

- **Low-Hanging Fruit**—These are opportunities that do not require data, can be found in every workplace, office, or workbench, and can be fixed using solutions readily visible to the workers. At the low-hanging fruit level, there is little or no data required to reach approximately 30% of the simple opportunities that reside in most well-performing organizations. When talking about low-hanging fruit, these opportunities are so obvious that specific Lean skills and learning are not required. Unique to this segment, there are "lessons learned" that are acquired by finding these opportunities and solving them simply and quickly. Frequently, they are on a lower level or smaller scale. The lessons learned gained as individuals or small teams can be as equally important as the more sophisticated or larger challenges because they can be communicated quickly and used immediately by individuals and small teams to find like-type issues and solve them.
- **Great "Waste"-land**—Opportunities abound in the middle ground; they are everywhere, usually involving more than one person or team, and data is present for simple analysis and solution generation. This is where the majority of waste resides in systems and processes implemented within most organizations. In the early roll-out, you will find approximately 80% of all work done, regardless of organization type or work type, is non-value add or waste. Data and analysis needs in this layer range from very simple to slightly complex. Traditionally, it is the

data used daily to run the operation and it contains more than enough content, or data richness, to be used for these analyses. Simple data analysis tools are generally used at this level including Pareto analysis, cause-and-effect diagrams, histograms, and other simple parametric analysis.

- **Tough Ground**—At the top of the tree there are opportunities, while impactful across larger landscapes of the organization, that are more complex and require greater data and more sophisticated analysis. The opportunities have confounded or often missing data and information, the problems are difficult to identify, there may be multiple root causes or they are difficult to identify, and the tools required for their analysis are more complex. Typical tools at this level include design of experiments, causal relationship analysis, two- or three-dimensional scatter diagrams, detailed statistical analysis, performance capability measures, and additional statistical tools.

Based on the zones of opportunity, a lot of waste can be found and eliminated in the early stages of transformation without the need for seasoned Lean practices, tools, or teams. This is goodness—it gives the organization the opportunity to make quick wins, share lessons learned through the activities, and begin to become more practiced and comfortable with the simple tools of Lean.

Between the low-hanging fruit and the great wasteland levels you can find and remove approximately 80% of inefficient, non-value added work. Most of this improvement can be obtained by individuals or by small Kaizen teams.

Obviously, having Lean tools and methods deployed throughout the organization helps foster more rapid harvesting of the waste. Only when the organization has been through several cycles of improvement and waste extraction focused on low-hanging fruit and great wasteland levels should you consider looking at the tough ground. More sophisticated analysis, higher levels of technical expertise, and more investment of time and resources are required to venture for the fruit at the top of the tree. Your job and that of your staff is to use good decision making to help determine what is critical for mission success, and prevent upside down investment in improvement actions.

5.4 Focus on Using the Voice of the Customer

> There is only one boss—the customer. And he can fire everybody in the organization from the chairman on down, simply by spending his money somewhere else.
>
> **Sam Walton**

Customers must be actively (not passively) pulled into the journey so their voice about satisfaction and dissatisfaction is clearly heard. The messages become a focusing engine that helps target where your most important partner in your success is heading, what they need, what they want (not necessarily the same thing), and how they judge your performance.

Are you listening to your customers? Have you heard where they want to be versus where they already have been?

The Voice of the Customer (VoC) is used as proxies for a range of markets the organization services. What is the VoC? Briefly defined for the purpose of these discussions, VoC is:

> Voice of the Customer is the system of logically segmenting customers into classes, intelligently asking for and receiving feedback, distilling and extracting key information and messages from the feedback, and focusing the information and feedback into significant guidance for changes to deliverables (product, service, and relationship) received by the customers.

A crucial concept that is often misused is the inputs from the customer must be used as a focusing engine during changes, not the driving force behind the changes. Changes must be coming from within the organization—its need to continuously improve overall performance. The VoC should help tune where, how, and what is changed.

- What is the voice of the customer?
- How do you listen?
- Should you be listening?
- What are you asking them?
- What are they telling you?
- How do convert what you hear into meaningful guidance as changes are required?

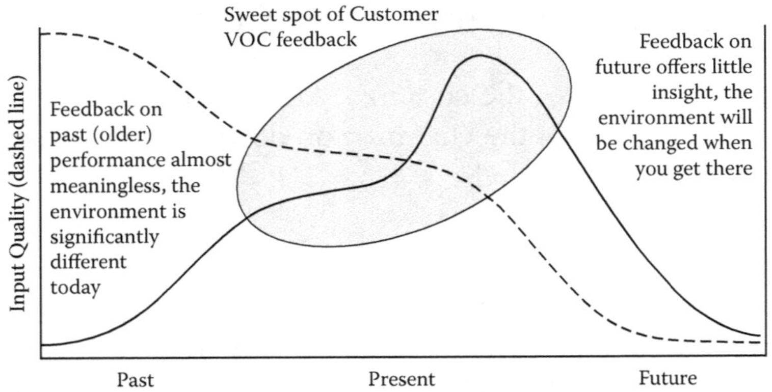

Figure 5.6 Listening to the Voice of the Customer.

These questions, and hundreds more, are the focus of this section. If there is a section that carries one of the highest impacts to changes you will make in your organization, this is it!

In Figure 5.6, an interesting picture emerges. Customers say many things about what their needs and requirements are today, where they are heading, what their future needs and requirements are, in what directions they are heading, and reflections of your performance, good or bad, as though they are looking in the rearview mirror of the path already traveled.

As you combine and integrate everything the customers say and then distribute the findings over the "landscape of relevance" to the organization, a somewhat familiar distribution emerges.

In the vast middle is all the information the customers provide about today's ongoing relationship, needs, and expectations. They usually root feedback in current products, services, or deliverables that are exchanged between your organization and the customer.

When looking to the left of the landscape, you find diminishing impact to the organization. This is especially true the farther back in history where the information is pointing or related. Frequently customers are happy to relate stories of difficult times that were important a long time ago, but are not relevant today because many things have changed since then. The customers will still try to use those stories to add leverage to their needs and expectations for today. It is important to be able to filter out what is meaningful and what is not.

Looking to the right of the landscape these messages are the signals the customers are sending about the directions that they are heading. Some of it is wishful thinking, some of it is desired state, and some of it is pure speculation. By using the same techniques but different filters, this information needs to be distilled into meaningful directions for the organization's focus on what it should be doing next.

5.4.1 VoC Segmentation

One of the initial actions most successful organizations do is segment their customer base. It is easy to recognize that the customer base is not monolithic, but is composed of discrete major classes of customers, each with their own needs, wants, and expectations. From your organization, some customers primarily receive services or support, while others receive almost entirely products or tangible deliverables. When you analyze the deliverables, it becomes evident that some classes of customers are driven by "commodity-like" needs while others require tailored and customized solutions.

> The more you engage with customers, the clearer things become and the easier it is to determine what you should be doing.
>
> **John Russell**

Why is this important? Because as the organization is driving through changes to its systems, processes, and delivery mechanisms, the concept of one size fits all does not work. There is a layer of complexity that must be interlaced into the changes to accommodate all classes in your customer base. Changing a process that provides a tailor-made solution to your customers may actually hurt the relationship with customers that only want simple, consistent commodity deliverables. The changes increased their work to order, or define, what they want, and it reduced their level of satisfaction. If you are not asking and listening to all classes of customers, there is a high likelihood that you will miss critical guidance as you drive for better performance.

In Table 5.2, you will see an example of a customer segmentation framework unique to a healthcare organization. It is divided into a matrix of customer segment rows, and expectations, needs, and success criteria columns. Contained in each intersection are the extracted messages, information, and success criteria unique to the customer segment and their needs.

While the constructs of Table 5.2 are sample segmentations with associated expectations, needs, and success criteria, what emerges are clear differences in the needs and expectations extracted from the different classes of customer segments. In order to critically gather customer information, there must be effective approaches that include both passive and active interactions—described as listening posts.

5.4.2 Customer Listening Posts

With the customer base segmented into logical classes most relevant to your organization, the next action is collecting feedback. The myriad approaches and methods are vast; compounding this with the array of kinds of customers and organizations, you end up with a total potential range of solutions that becomes so large that you cannot count the ways.

As noted in the book *Keeping Your Customers* (Lowenstein, 1995), the author cites studies that customers who do not provide negative feedback are usually anywhere from ambivalent to mildly upset with their customer experience. He demonstrates that there is a threshold of negative perception a customer will tolerate before making an overt action and contacting an organization to give it feedback. However, by then it is too late—the customers are highly negative, usually with several examples that prove they did not get the experience they wanted and the most recent one is the one that broke the camel's back. Without active listening posts, you never knew they were having negative experiences, let alone enough of them to negatively affect their future repurchase decisions.

Now consider that you have active listening posts and passive listening posts, each serving a distinct and appropriate purpose to capture customer feedback. Some examples for each are:

- Active listening posts
  - Outbound customer surveys at targeted customers using customer-contact persons
  - Service completion polls done using human intervention
  - Point of sales questions to determine satisfaction
  - Active customer feedback meeting
  - Product or service customer tests
- Passive listening posts
  - Comment and suggestion collection points
  - Surveys that require an overt act initiated by the customer to engage and complete a set of inputs
  - Third-party scans of social media comments

Where do you go with this? Let us continue using the healthcare customer segmentation table example (Table 5.2). You can establish segments that require feedback by creating a quick listening post table. In Table 5.3, the unique listening posts for each customer segment are noted in the segmentation example. Some of the methods used by different segments are the same or similar in construction and use. Through disciplined collection methods a greater depth of customer knowledge is obtained and can be used more effectively as system, processes, services, products, or deliverables are improved or provided.

By now, you understand your VoC segments, have active and passive listening posts in place, and receive data and information from customers. Questions naturally emerge:

- What do you do with it?
- How can you leverage what the customer is saying to your advantage?
- Are there signs that you are winning or losing the race to embedding a sustainable competitive advantage?

The next topic will explore how you can use the VoC as a focusing engine to extract a laser-like view of what is required for success in the customer's eyes.

5.4.3 VoC as a Focusing Engine

Why is VoC called a focusing engine and not a driving engine? The concept of gathering and using the VoC to be able to provide valuable, and frequently unfiltered, feedback is critical to being able to see traits not being performed well or being performed in a way that customers do not see the benefit.

The dichotomy of this approach is that changes to systems and processes that affect the timeliness, quality, and price of the deliverable should be coming from within the organization—from a culture that is continuously dissatisfied with current levels of performance and is striving to learn, grow, and improve. If an organization waits until the VoC feedback is received and then it reacts, it is too late. The customer has moved on to the next provider and future customers are now being tainted by the defected customers as they relate their experiences and stories on why they did not stay as a customer.

This implies that changes, both ad-hoc and planned, should be continuously in flight, and the cycles of improvement should be aggressively seeking and using VoC to tune the new methods, products, or services.

Table 5.2 VoC Segmentation, Healthcare Example

Segment	*Expectations*	*Needs*	*Success Criteria*
Patients	• Safe, effective, timely care • Emotional support • Family involvement • Comfort • Communication	• Appointment notices • Post-discharge communication • Patient rights information • Complaint process	• Healthcare outcomes • Patient advocacy data • Patient loyalty data • Readmission rates
Families	• Family involvement • Effective communication • Information for supportive care	• Post-procedure follow-up calls • Post-discharge instructions • Communication • Access to healthcare providers for follow-up • Focus groups as needed	• Call wait times • Healthcare outcomes • Family support post discharge • Family satisfaction data
Partners	• Communication • Consultation • Diagnostics • Recognition • Information • Collaboration	• Communication • Outreach events • Electronic bidding and payor processes • Partnership services updates	• Healthcare outcomes • Outreach and collaboration events • Communication capabilities • Payor process effectiveness and efficiency
Regulatory organizations	• Regulatory and accreditation compliance • Clear communication • Concern resolution • Timely feedback	• Governance communications • Regulatory correspondence • Audit and evaluation hosting • Timely compliance submissions	• Compliance and governance requirements met • Successful audits and evaluations • Audit findings • Findings resolution

Table 5.3 VoC Listening Posts, Healthcare Example

Customer Segment	*Listening Posts*
Patients	Complaint process, discharge follow-up conversations, formal surveys, informal feedback, patient interviews
Families	Clinic visits, complaint process, focus groups, patient advocacy forums, formal surveys
Partners	Advisory committees, governmental agencies, community events, complaint process, patient interviews
Regulatory organizations	Accrediting agencies communication, community events, complaint process, statutory regulations, interviews, regulatory agencies communication

In Figure 5.7, a simplified model shows a VoC system that includes VoC listening posts to gather data, raw information, and anecdotes from customers feeding a filtering and distilling process that separates the wheat from the chaff, and then aligns the final messages using the customer segments. These customer-segmented messages are then used to focus improvement plans already planned, or in flight, so the ending solution is best tailored to identified customer needs or expectations.

An additional element emerges as listening posts gather raw comments and information from customers—a simple measurement of their satisfaction. Traditionally, collected inputs have two attributes scored by the customers:

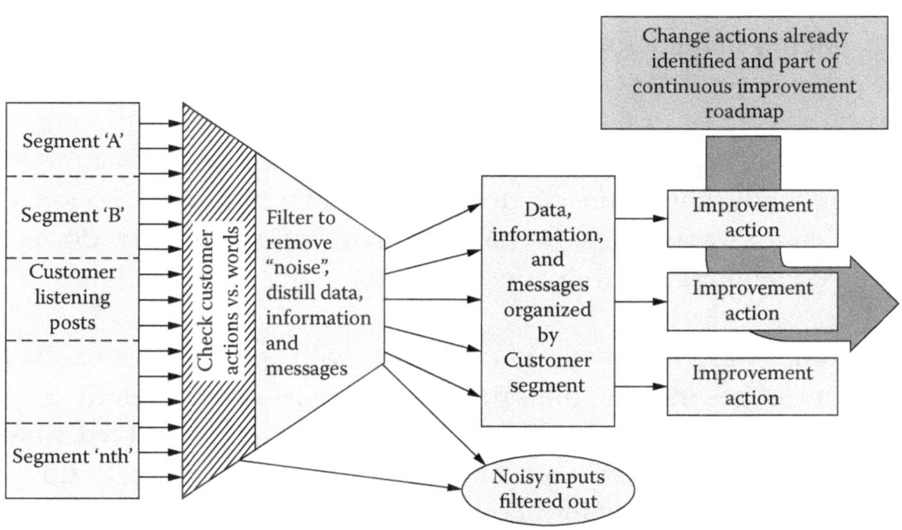

Figure 5.7 VoC listening, segmentation, and improvement focusing flow.

- Importance of the deliverable to them, and
- Satisfaction with what has been delivered.

Effective filtering does not require complicated analytic tools, but it is specific and unique to each organization, requiring tailored methods to separate the wheat from the chaff. However, the following example shows how customer inputs, while valid at the micro level or at the specific input level, are ultimately not valid because their actions do not support the request. The example comes from a technology-based manufacturer for a large high technology customer.

> For several years, a major customer of a high-tech organization asked for cost concessions for the task of breaking bulk-pack products down to single package items sellable as units of one. The high-tech organization used a third-party distributor to perform the breakdown tasks and charged the customer for the service. The organization responded to the customer without reducing the cost for bulk repackaging services. After the customer made the request for several years and received the same response each time, the organization finally proposed to the customer that it could easily contract with the same third-party distributor to do the work directly for it and that way it could reduce the single-package item cost independently, on its own terms. The customer did not accept the proposal and continued to ask for concessions. That response was offered multiple times; in fact, every time the customer made the same request for cost concessions. After the request cycle repeated itself several more times, it was clear that the customer would not accept the proposal and solve its issue on its own. Consequently, any future requests of the same type were no longer considered as reasonable or viable in terms of customer feedback and were eliminated from consideration.

This example shows how good customers with good intents can insert disruptive messages into the important customer feedback system and cause significant consumption of time and resources to respond and react because what they are asking for does not match what they are willing to do.

By integrating the scores into Equation (5.1), you end up with a gauge of how your customers rate you, especially targeting the middle scores where customers are neither satisfied nor dissatisfied.

$$Customer\ Satisfaction = \int_{1}^{n} Customer\ Experiences \tag{5.1}$$

where:

PVMP = sum of negative experiences for incidents with individual customers

The concept of PVMP is best shown in Figure 5.8. The overall distribution of what is being delivered to the customer, the entire distribution curve, is usually measured by the average of a critical expectation of the customer. In the figure, at the far left edge of the original distribution denoted by the dashed line, is an area that missed customer expectations. This small area has the ability to taint the overall customer experience—it is the origin of the PVMP.

Why is this important? Because when you apply Lean techniques to a process, almost any process, an immediate reduction of the variation of the output begins to be realized. Additionally, the learnings gained through the examination, root cause analysis (RCA), and determination of how to better control the process can be quickly shared with other process owners in an attempt to accelerate their performance results.

In Figure 5.8, this reduction in variation is seen as the distribution marked with the solid line. You can see where the PVMP zone has been eliminated through reducing the variation of the distribution of what is delivered to the customers. Notice that the center of the distribution has not changed but the variation is now much tighter. Additionally, the full distribution is better than

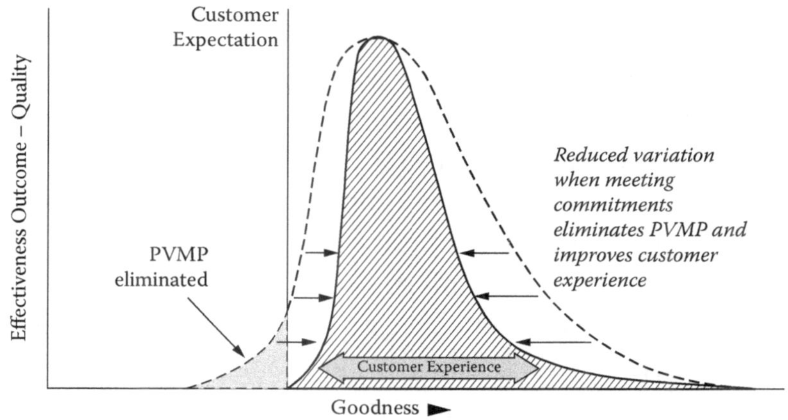

Figure 5.8 PVMP eliminated via reduction in variation.

customer expectations. The area outside customer expectations can no longer taint the customer experience. While this is an example, the distribution is typical for single-sided customer requirements (i.e., on-time delivery must be at, or above, a given threshold).

The key message is that the PVMP is founded in the statistical distributions of a process, and that through process analysis, evaluation, and tuning you can quantify statistically the impact on the customer's expectation for a given performance level. You can also calculate how impactful a process change would be to the customer experience. Given the value of resources needed for improvement, you can tell how much it is going to cost to get a given change in the customer's satisfaction with the organization.

5.5 Setting Simple Goals

> Productivity is never an accident. It is always the result of a commitment to excellence, intelligent planning, and focused effort.
>
> **Paul J. Meyer**

The innovators/explorers and early adopters are off making impacts to the organization; now it is time to get the other 80+% of the organization on board and focused. Here again, focus will help in convincing the people that they are capable of having an impact and they need to commit to the strategy.

There is no better way of doing this than setting simple, achievable goals, which are looking for demonstrated practice. You want to see examples that people have tried to use the tools in a way to achieve the desired results. In some cases, this will be process improvements, working in a group environment on a Lean project, taking part in Kaizen events, etc.

Earlier, we commented that it was simple to find projects to focus on. The easiest way is to simply ask an individual what particular aspect of his or her job annoys him or her the most. By default, people understand what does not make sense or is not effective or efficient. By asking the Five Why's and by focusing them on the portion of the problem that they control, you can get anyone started on the track of improvement.

For those people who need to have a quantifiable objective, there is nothing simpler than saying, "Deliver twice the good, or half the bad."

The objective is to get people thinking. What is simpler than saying, "Do something twice as good," or "Improve process performance by 50%"? You cannot get that level of improvement simply through increased effort.

You have to address the way the work is done. In doing this, it will quickly become apparent if they are eliminating steps, waste, touch points, opportunities of defect, or merely squeezing the gas in the balloon. If the improvement that someone makes simply transfers the workload to another part of the organization, internal or external, you realize it is a false improvement effort. Where someone transfers the work from a manual step into an automated step without first simplifying the process, you realize there is false savings. There will be many false steps, so you have to be capable of identifying why the improvement that someone thought he or she delivered was in fact not what you were looking for. This feedback mechanism is critical for the individuals to start to understand what the application looks like through the lens of elimination of waste.

As you consider the multiple dimensions, or functions, in which you are trying to move this initiative, it is important now to consider all aspects of the hierarchy within the organization. Your team and their direct reports in the managerial ranks will be evolving at different rates as well. The deployment of management standard work, which we will describe in more detail later, will become essential in your ability to communicate and articulate in a common language across the group methodologies that span very divergent organization processes. This in effect will become the vehicle for your management team to "walk the talk."

Finally, leveraging peer pressure by having managers report status of activities and initiatives in staff meetings allows you to get a sense of their level of understanding and comfort with the material. Focusing on small projects and turning them quickly is a real plus. It builds personal cash for the Lean drive quickly.

Resist the organization's natural drive to want to solve the big, hairy issues with Lean. They are often complex and require other considerations. By winning at the small challenges, you are unwinding the complexity of the big issues. There will be a natural time when a large issue is ready to unravel and be fixed.

For the leader, tie this to the early rising slope of the deployment curve—it takes a lot of energy to overcome the hump.

- Get uncomfortable with the status quo.
- Do not over-design the end-point, direction, or vector.

- Ensure enough credibility to set the direction and course of action.
- Be open to random and structured improvement techniques; waste is everywhere—it is okay to grab low-hanging fruit.

Staying constantly enthused with transformation and change without causing cultural disruption is one of the challenges faced by the senior leadership of an organization.

5.5.1 Work on Things That Matter

> There is nothing so useless as doing efficiently that which should not be done at all.
>
> **Peter F. Drucker**

LWL is not the pursuit of perfection. One form of waste is designing processes or capabilities beyond what is material or meaningful to the customer or organization. It therefore becomes critical to have a process in place that allows you to understand where the Pareto of opportunities lies.

Driving process improvements in one area may consume valuable resources that would net much more significant benefits by working on a different element in the process. This again represents another dichotomy with LWL. You are looking for an activity in levels of engagement across each function within the organization, but at the same time are trying to provide focus and direction to ensure that their efforts produce tangible benefits. It is easy for an individual to stack things he or she wants to address in priority. Similarly, managers can typically identify the sequence by which they want to address improvements.

Your role is to look across functions and help them define those areas where they should be investing time and energy. Having value streams for the critical processes in place with a Pareto of causes for deviations of performances and an action register that identifies the next opportunity tackle will help keep the organization focused.

Work on opportunities that matter to the organization—some small, some large. The challenge is to set the prioritization method in place and use it to determine what to work on (you only have so much improvement energy—spend it wisely).

Work on stuff that matters to the people, too. If you make their work easier and less disruptive, then they will be more likely to accept and adopt the new ways of the transformation journey.

5.6 Keep It Simple

> Simplicity is the ultimate sophistication.
>
> **Leonardo da Vinci**

Always use the KISS (keep it simple stupid) principle for process improvement. This is easily said and hard to do. The more technical the environment you work in, the more likely you are to have people who are trained to work with complex ideas, systems, and problems. Getting them to realize the beauty of simplicity and to start construction of solutions with that in mind is no small task. Learning to see the waste is critical for the general employee population.

Some common evidence when KISS does not get used includes:

- product or process over-design
- projects with scopes being too large
- identifying opportunities for improvement outside your immediate sphere of influence or control
- identifying the areas where others can improve
- not having clear benchmark data to indicate if the results achieved represent an improvement over similarly improving BIC processes

There are numerous examples where over-design and complexity failed to produce the desired results: the cost and energy that NASA expended on developing a pen to write in zero gravity compared with the simple solution the Russian space program came up with for the cosmonauts—a pencil.

So, too, with the other pitfalls listed previously, it is clear that what the team worked on is almost as important as the learning gained in the exercise. This book is not trying, yet again, to write an instructional manual on Lean principles. It is important, however, to recognize common mistakes as you deploy LWL throughout the organization.

Avoid temptation to automate manual processes. The focus here clearly has to be on simplification and process improvement before anything is automated or hardcoded. The concept of micro-waste was discussed earlier and this is a prime example of how it creeps into processes. Another simple error is to identify how others can improve without looking at what we control ourselves. It is still easy for an organization to gravitate to roles, responsibilities, and historic practices when thinking about improvement.

You are trying to get people to recognize what the real value add is and work on simplifying and improving that element in the process. For that focus to take place, it is best to have the team first examine what they actually do to support the end-product. Seeing the waste in replicating data in multiple emails, replicating the work necessary to start an activity serially instead of having prepared in advance, and recognizing the number of times a specific task is started, interrupted, restarted, and interrupted again are valuable ways that the team can begin to see the waste in the process.

> It seems that perfection is reached not when there is nothing left to add, but when there is nothing left to take away.
>
> **Antoine de Saint Exupéry**

If you cannot do it with a pencil and a piece of paper, it is too complex. Notwithstanding the necessity of complex data mining, automation is the last thing you want to do during process redesign.

The following unattributed story is about a simple solution that brings home the point about listening to solutions from the workers who know the process, regardless of your industry or organization's deliverables:

> The General Manager of a successful production facility that filled squeeze-tubes with a product, and packaged each tube individually in a small box, noticed a sustained, low-level trend of empty boxes making it to the end of the line. This prevented the facility from achieving 100% goal attainment. The GM went down onto the floor and saw the problem first-hand. Wanting the production line to achieve its goal, the GM commissioned a consulting firm to design a solution to assure 100% quality for boxes containing a tube of product. After kicking off the project, 8 months of development with several solution reviews with senior managers, and $500,000 invested, they implemented an automated solution that detected an empty box, sounded an audible alarm, and ejected the empty box off the production line.

The GM was initially satisfied because the defect rate dropped to 0 (no empty boxes were making it to the end of the line). Seeing the results, he again went down onto the floor to watch it in action, and discovered, to his surprise, that no empty boxes were being found or ejected by the newly installed solution.

Confused, he asked the operators why no empty boxes were being found. He could not figure out what had changed in the process because previously there were empty boxes escaping all the way through the production process, and there were none now. The operators quickly answered that the new solution had a noisy alarm every time it found an empty box, and it required someone to walk over and turn it off each time. So, they put a small desk fan farther upstream that blew across the conveyer line. It blew empty boxes off the conveyer line before they got to the new automated (and expensive) solution. Their solution to the problem was so simple (and cheap). It meant they did not have to hear the alarm anymore, or get up and go over to turn it off. It saved them time and annoyance and it still assured 100% full packages at the end of the line.

What is the message? Learn how to keep things simple by working with those who know how it works, and can fix it simply, quickly, and cheaply. For the cost of a $10 fan and 10 minutes worth of work, they had a solution that was assured 100% quality and was far better than the sophisticated design (far cheaper to install, faster to implement, and cheaper to maintain).

5.6.1 One-Touch Flow

A cultural behavior underlies the concept of OTF. It revolves around the practitioner's view of how processes work, should work, and contain various kinds of waste. Imagine if you will, every time you walk on the floor you look at work being done and you see the inefficiency, ineffectiveness, or variability in its execution. One of the key troubles with having the ability to see OTF is that you are always frustrated because you cannot understand why others do not see what you see. Some call it a gift; others call it a skill. In any case, is a critical capability that good Lean leaders in Lean champions need to have.

OTF is crucial at this stage; it provides one of the philosophical foundations for easy process simplification. The following questions begin to set the context for OTF practices:

- Does everyone understand the quality standard that the person receiving the work will use to validate that it is acceptable?
- Does everyone understand the advantage of touching a piece of work once through to completion at the acceptable quality level without interruption or rework?
- Are there clear visual indicators of capacity utilization at each point that value add is done (including your desk)?
- Are you measuring what actually is changed, not what is supposed to change because of process improvement or modifications to a design?
- How do you know when a decision to reprioritize activities is being made?
- Does everyone on the team understand who his or her internal customers are?

These simple checks will help you focus the group on areas for improvement and will provide many opportunities for deploying Lean principles.

Another important expectation of the organization is that time and speed are essential to successful performance; this means you cannot stop critical processes, go through cycles of improvement, and then turn them back on. They must be improved while they are in flight, and solutions validated transparently, and then cut over to replace the earlier problematic method. The concept of improvement on the fly captures this point.

Technology has the ability to make processes more wasteful especially when it is easier to automate a poor system because it is faster to let technology do the extra steps instead of taking the time to remove the extra steps and make technology run a streamlined process.

There is a basic equation on process simplicity—fewer touch points equals an automatic improvement in process quality and attending first pass throughput yield without improving each incremental step's output quality.

In Figure 5.9, you see a natural improvement in quality through the simplification and elimination of many steps into a few remaining steps. The simple equation that defines the quality for a process like this is a serial probability function. It becomes the multiplicative accumulation of the probability of success of each of the individual steps. As you can see when the probabilities are multiplied, the resulting quality is significantly less than when just a few steps are multiplied together.

Let us use the example in Figure 5.9. In the top half, it depicts a seven-step process, each step having a unique error (failure) rate. The probability of success is the serial failure rate (i.e., Task 1 has a 3% error rate [0.03] that equals 1 – 0.03 to give 0.97 probability of success, times Step 2 failure rate of

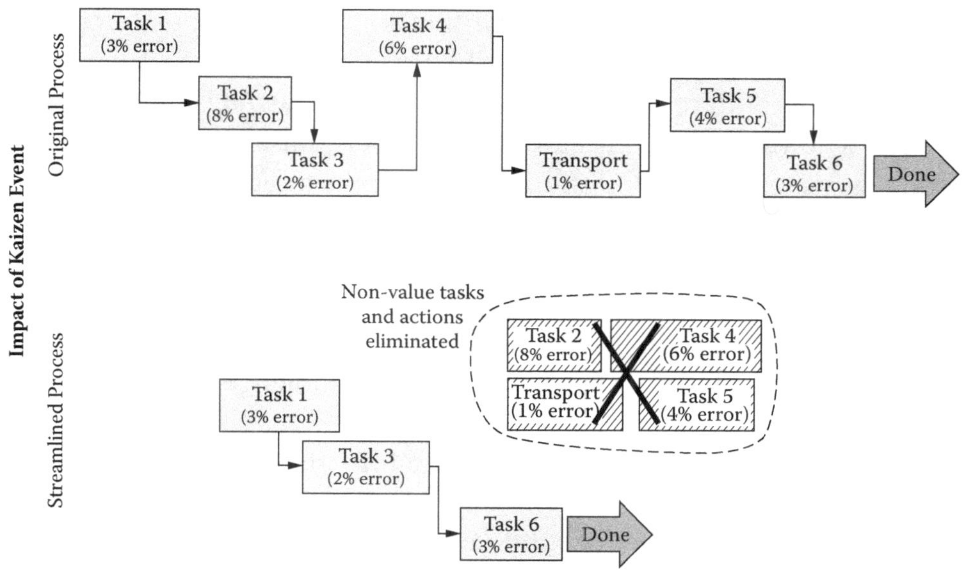

Figure 5.9 Inherent quality improvement, original versus minimalized process flow.

8% [0.08] to give 0.97 × 0.92 = 0.89, and so on down the line to the last step). When all seven steps failure rates are multiplied together, you get the overall system's probability of success (i.e., 0.97 × 0.92 × 0.98 × 0.94 × 0.99 × 0.96 × 0.97 = 0.76, or 76% probability of success). It is relatively easy to understand that being successful approximately 3 out of 4 times (76%) in the eyes of the customer is likely to affect their repurchase decision making.

Now, use the bottom half of Figure 5.9 to see that four of the seven steps have been eliminated. Immediate gains in throughput can be seen, but at what improvement in quality? Using the same logic, only the remaining three steps are calculated into the final value (i.e., 0.97 × 0.98 × 0.97 = 0.92 or 92% probability of success). This represents a gain of 16% in output quality by simply removing steps. The quality of each step was unchanged, only steps were eliminated. The inherent value of OTF becomes evident in this example.

5.7 Communicating to the Organization

Regardless of the changes in technology, the market for well-crafted messages will always have an audience.

Steve Burnett

This chapter is all about Phase 4 of the transformation cycle—Doing and Improving Processes. This is accomplished by using what you have learned to accelerate future improvements and even faster learning. By now you have been leading the organization through four phases that include investment in assessing and finding strengths and opportunities, planning a large transformation, up-skilling all employees with new improvement and learning, and then using the new skills to improve a wide range of opportunities. You have communicated with many different audiences ranging from the entire population down to targeted, small Kaizen teams. You have had to porpoise content and delivery based on your audience and which phase they were in.

In this phase, a few significant topics are important to cover:

- Reminders of what the goal and end game are supposed to look like.
- The picture of growth and change from where you were when you first started to where you are at today in practicing good organizational learning. How much closer to a change-hungry organization it is, and how you see evidence of having a sustainable competitive advantage.
- A summary of the operational impacts that the transformation has had for the organization—a change in culture, behaviors, and actions; demonstrating new techniques, tools, and skills; and changes in performance internally and externally.
- A reflection of what the organization has learned, what the management team has learned, and personally what you have learned throughout the experience.

Additionally, you should seize the opportunity to help the organization understand and translate the reflections you just offered and how they need to be accelerated and continued moving forward. This is not a one-time shot; it is a journey that does not have an end—only mile markers and milestones.

Key Messages in This Chapter

1. Improvements must be self-initiated at the individual contributor level, your early adopters and Lean SMEs can't be the only source of these activities.

2. There are two basic approaches to improvement—attack the outliers in the process (special causes) or improve the process capabilities (common causes). You need to do both, but they will typically have different impacts and resource requirements. Focus on special causes can often generate quick wins.
3. Every expenditure of energy in this transformation has to be assessed against how it improves the business, improves the rate of learning, increases the problem solving capabilities of the group or adds to the combine CCI.
4. The decision to use continuous improvement or Kaizen events is less important than the decision to start doing something that helps the organization to learn using lean!
5. We recommend the 3 steps for process development: 1) process improvement, 2) process control and capability management, 3) process design. Getting these in the wrong order can consume a lot of resources for questionable results.
6. The higher you are in the organization, the farther you are from the process, the less you know how it works. Get the people that do the work to make the improvements.
7. Embrace your ignorance and learn from those around you. Learn to listen, and see how things are done and improved. As your learning curve increases so will your expectations for the organization.
8. Don't automate a "noisy" or waste ladened process.
9. Over use of data analysis can paralyze your efforts…get good at try-storming…you at trying to get the organization used to the notion of learning how to deliver structured improvements. Perfection is not required! Repetition is required!
10. Use your customer feedback in the correct context from planning horizon perspective…Input on data that is too old or based too far in the future is of little value. Know the sweat spot for your organization.
11. One simply way of communicating your intent is "through lean we will teach every element of the organization how to reduce the PVMP!"
12. One of the toughest elements of the journey is to focus the team on fixing the simple things as a means of addressing improvements vs. getting lost of the ornate/complex big bang approach to problem solving.
13. Use a critical eye on every process to identify if One-Touch Flow is evident.
14. Start bragging about the step function improvements the team is delivering. Your excitement will be contagious!

Challenge Actions

This section has been about Go Improve Something—Start Doing. It is all about deploying the methods and changes to the organization. Lean improvements should be evident in most areas and the language should be well understood by this stage.

Actions now focus on building the organization's "muscle memory" in improvement practices rooted in Lean techniques. Critical to keeping focus and attention to details, routine management reviews of progress and results are essential. Select enough actions below to total 50 points.

Challenge Actions – Go Improve Something – Start Doing	*Points*
1. Conduct a simple VoC exercise by focusing on the top five customers. Call them and ask a series of structured questions, gather their feedback, and distill the feedback into meaningful information. Evaluate your listening posts and customer segmentation for sufficiency to organizational needs.	25
2. Take a personal walking tour of an area that has processes directly associated with your organization's core deliverables. Afterward, record all the processes you observed and your impressions of their effectiveness, efficiency, and consistency.	15
3. Re-evaluate the CCI of you and your management team. Cross-compare and identify natural leaders in each segment and establish a mentoring approach to leadership development.	10
4. Compare and contrast management standard work (MSW) as it is deployed across the organization. Identify strengths and institutionalize them. Identify where there are still gaps and build "next steps" action plans to help close the gaps.	20
5. Which 5 of your managers show evidence of their competence in Lean tools and methods by conducting simple tests of "watch one, do one, teach one."	15
6. Look for evidence of standup meetings that are working well. Capture the lessons and share organization-wide.	10
7. Find a self-directed team, attend some of their improvement meetings and capture what they are doing well. Then, discuss with their managers what it took to be assured that the team was capable of effective self-management. At your next staff meeting, share the findings and develop a plan to propagate successful techniques.	20

8. Introduce the subject of learning into each staff meeting, with each member presenting things he or she has learned in his or her own deployment of MSW or other Lean practices	15
9. Create a standing excuse to take a walking tour each month, or more frequently if you travel to other locations, of localized work areas. The purpose is to show the organization that you are actively trying to break down the insulation at the top of the management chain.	10
10. Have a senior staff member identify an automated process that is problematic (or wasteful in its labored execution) and lead a discovery team to find at the lowest detail level (stepwise) where automation only made a bad process worse. Capture the decision-making process that drove the implementation for reference.	25
11. Have you created an environment of shared learnings? Do you have time for staff members or single contributors to share the results of projects where significant improvements have been made?	25
12. How many of your direct reports are using VSM? What percentage of your processes is covered by VSM?	

Chapter 6

Leverage the Learning

> Once you climb to another level, you have to figure out how to sustain it.
>
> **Mary J. Blige, 1971–Present**
> *Singer, Songwriter, Producer*

The journey in the first four phases of your transformation led you from assessment through planning and execution, all the way through to finding and eliminating wastes embedded in most processes within an organization. It is a job well done, but not complete. How do you leverage what you have learned? Internally? In your supply chain? With your customers?

Most organizations can reengineer a process, but is your organization learning how to learn? Leadership organizations are the ones who develop a hunger to change from within, to constantly learn from their successes and mistakes. This fundamental philosophy has to be built into the way an organization drives itself to higher levels of performance.

In this book, the focus is on building competencies and capabilities for sustainable competitive advantage. This requires a fundamental shift in the way the organization looks and acts on challenges and opportunities. Many examples and discussions focused on the idea of using Lean tools and methods to drive organizational learning that achieves a Lean Learning Organization culture. This is where Lean is woven into the organization, its daily actions and behaviors, and it promotes and enables the behaviors of a Learning Organization.

A concept worth refreshing is the difference between process reengineering and being a Lean-based change-hungry organization. Any organization can reengineer its processes and systems. While the target is for the better, faster, and cheaper systems, frequently the opposite occurs. This is good, but not good enough. Being a change-hungry organization, using Lean to achieve organizational learning and focus, and driving to obtain or achieve a sustainable competitive advantage are the end game. Change-hungry means the employees and the organization are constantly digging and finding more efficient and more effective ways to transform and deliver to the customer's expectations.

Built into this concept is the framework of learning to learn. Successful organizations are frequently mired in legacy systems, historic methods and actions, and procedures and policies that are not as dynamic as the environment in which the organization survives. Process reengineering does not address these issues; it has a very microscopic approach at the product, service, or deliverable level.

Learning to learn quickly brings to mind the old saying "You can't teach an old dog new tricks." The implication is that an organization is hard set in its ways and incapable of change. Vibrant, healthy, and productive organizations have the ability to make this patently false.

A suggested tactic may include collecting and sharing the lessons acquired in the organization with external bodies that provide peer and competitive review to determine the level of maturity and excellence. Whether there is recognition through winning or not winning an award, the feedback processes are crucial to be able to understand and evaluate opportunities and elevate to the next levels of performance. Examples include competitive processes like:

- U.S. Department of Commerce Performance Excellence Award (Malcolm Baldrige National Quality Award)
- Shingo Prize from the University of Utah (Lean Performance Excellence)
- Deming Prize (Japan's Performance Excellence Program sponsored by the Japanese Union of Scientists and Engineers)
- Competitive and certification programs in many other countries, regions, and localities that may be unique to certain industries, such as education, healthcare, and public and nonprofit sectors

In Figure 6.1, a unique sub-process identified as try-storming resides in the transition between Phase 4 and Phase 5. In practical terms, this is a small mini-cycle of trying something, examining how well it succeeds or fails, capturing the lessons learned, and redoing the solution in another attempt. The loop stops when the solution permanently solves the problem,

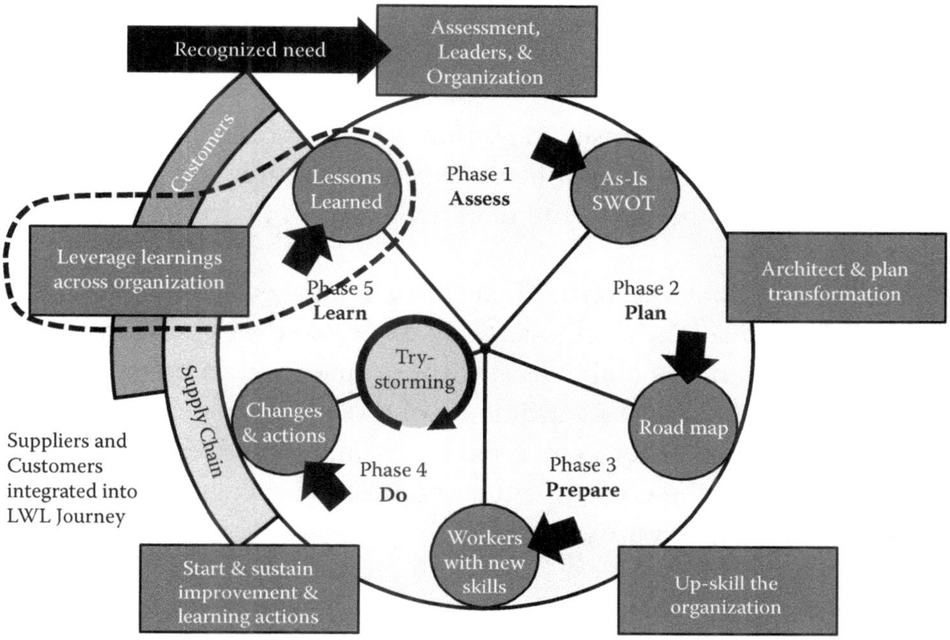

Figure 6.1 Learning, Phase 5—transformation.

is effective in how it works, and is efficient in the resources required for its execution. Try-storming is intended to maximize team involvement, knowledge sharing, solution creation, and permanently fixing the opportunity. If you will, the small loop is simply a consolidation, or micro-execution, of the larger transformation model.

6.1 Reevaluate and Refresh

> Faith in the endgame helps you live through the months or years of buildup.
>
> **Jim Collins**

As with any change initiative that is large in scale, you will go through an initial wave of deployment. After that first round of deployment, you must reevaluate and reassess to ensure it met expectations and, where it did not, go back and refresh.

Refreshing can take many forms:

- additional training
- changes in roles, responsibilities, authorities, and accountabilities
- growth in the change network of Lean and learning leaders
- changes or adjustments to the direction, course, or speed as needed

Early in the book, it was stated that this is a journey without end; this is where that hits home. Early on, questions were asked of your commitment to the transformation, the ability to stay the course, and to be the lightning rod for deployment moves forward. Each of those questions expected a "yes" response. This is where you have to dig in, reach deep, and add that extra little bit to stay the course so the organization recognizes the long-term institutionalization that you expect.

> To improve is to change. To be perfect is to change often.
>
> **Winston Churchill**

You can be assured that you will go back and reuse many of the tools that have been explored and their use already started. The tools are designed to provide relevant, new data and information, as well as provide trends from instance to instance of their use. This is perhaps the more important measure—the rate of change in the organization.

At this point, your expectations are to see positive trending in key measures. Do not be surprised if, in some areas, the key measures remain flat or even recede slightly. Those are important signals for you and your leadership team. They provide insight on how the selected plans and deployments may have missed the mark. Even with the best knowledge, foresight, and perceptions of the designers and architects in the planning phase, they are not always right.

At this point, you may discover that the organization structure and workflow do not follow a traditional model. Many organizations are functionally aligned, but in many cases, that is not how the work is done. Cross-functional work cells, or teams, may be naturally forming as the teams address ways to further simplify the flow. In some cases, you will have to invest some energy in this area. There will be a natural reluctance of the functional leaders to release their immediate responsibility for allocating resources as the team self-manages situations.

You will need to evolve the criteria that you use for evaluating your staff as this happens. The lines of performance ownership will blur. Assessing contributions will be more difficult. In fact, if you do not see evidence of this evolution, you just identified an opportunity for improvement.

Subject matter expertise is often a criterion for selecting managers, but in the new paradigm, this will only be one of the criteria for their success. The ability to put all of the opportunities into context and to help the teams select the path of improvement (project selection) will be a critical skill. Leveraging and assessing resources outside of your domain expertise will become a common task that some will find hard to accommodate. Managers who use position, knowledge, or seniority as leverage will find this change very challenging. Be prepared that some of these managers will not make the transition to the new modus operandi.

As you consider the evolution of self-directed work teams, consider that there are many possible origins from which they could evolve. If there is an obvious continuous flow of activities, the natural customers and suppliers in these processes may congregate to better align steps required and manage the flow of materials. This would be the traditional work cell approach as seen in manufacturing.

Another example of an evolution could occur when various functions recognize that the quality improvement projects they have underway overlap. In this discovery, the team will see that there are causal elements of the work they do that affect downstream processes. They can then start to address simple approaches to managing the flow in real time. An example of this might be the linkage between finance, customer care, supply planning, and commit, as it relates to order processing. The quality of the incoming purchase order from the customer is sometimes of an unacceptable level with critical data fields either incorrect or missing. Finance has the obligation to validate that the customer maintains an acceptable credit rating with the organization for the order to be accepted. There are many attributes relative to the product ordered that need to be confirmed as they relate to the specific customer demand. Finally, the question of availability, or security of supply, determines how you can schedule the deliveries. From this example, you could easily extend the steps to include logistics, customs, billing, invoicing, etc. as examples of how this particular stream connects. From this example, the people involved with this process would most likely work in different functional groups, with different organization structures, and often in different parts of the organization.

So how could this evolve into functional work cells or a cross-functional team? Once you have established an approach for the organization to share Lean projects and improvement actions that are happening, there is an opportunity for the extended group to recognize that efforts and energies on quality improvements that one team is making could have direct impact on the next function in the line. From this awareness, the group may decide that establishing a simple, 15-minute standup meeting daily to review all new orders from this customer may be the easiest way to capture the defects, resolve the issues, and identify the actions in a timely fashion, as opposed to doing it serially, over the course of days, using standard routings and email trails. While the causes of the defects need to be addressed, the turn-around time (TAT) for deploying this type of process is greatly reduced over the traditional method. What historically could have taken days or weeks to get a rejected order to the point where it could ship can now be handled in minutes or hours.

Another example of work cell evolution can be found in project-based work. A traditional approach is for various functions to perform their specific tasks and meet with some regularity to review progress. It is common to see schedules slip, as the delays in one stage of the project, by one function, move down the entire timeline.

By forming a cross-functional team leveraging typical work cell practices (e.g., single piece flow, line stoppage at detection of a defect, workload balancing by flow), the overall efficiencies of the team's outputs are maximized. Work is not done on the wrong version of the design, workloads can be balanced at critical gating items, and schedules can be compressed to the true constraints versus having schedule slips randomly cascade through the project.

However they occur, the presence and maturity of self-managed teams is a critical element in evaluating the overall progress of the organization. The closer the decision making gets to the point where the work is done, the more likely you are to avoiding over-production wastes. The better the group gets at assessing their performance on the last project or improvement activity (assessing their rate of learning), the steeper your rate of overall progress will be in LWL

The myriad of twists and turns available will ensure that no two individuals or organizations will develop in the exact same manner. This reinforces the premise that stated that LWL had to be developed from within and could not be purchased from a consulting organization.

A significant part of the simplicity of the approach of LWL reflects on its objective to openly learn from failures and successes. It is likely that you will have some of both at this point. Do not downplay the failures, openly share what you see and find with the organization. Treat problems as advantages, not disadvantages. Every time there is a problem, it allows creative people to create solutions. This achieves positive, measureable impacts to the organization and it helps bring the employees closer together. As their maturity rises, and through the use of a single vocabulary and simple, common tools, they begin to look and solve system-level problems, not just process-level problems. Remember that what you are looking at—in either positive, flat, or negative trends—belongs to the organization. It has the obligation and accountability to help identify root causes for poor performance and develop solutions for their remedy.

The spider diagram in Chapter 4 (Figure 4.3) can help you to visually assess the progress made on the various core processes for your team. The strengths and areas requiring additional investment will be obvious. By now, you know who the natural transformation leaders are and you can engage some of these team members in cross-functional projects to help accelerate the rate of learning in other areas.

By now, you have invested time to be able to transform into a Lean Learning Organization. The premise was that if you build those capabilities in your organization, practice them daily, and measure their impacts with feedback, then you would be building a competitive advantage. Even with more work to be done, there should be very pronounced improvements on many dimensions of the organization's deliverables.

Recognizing this, and the concept that the journey is never-ending, you still have to answer fundamental questions, such as:

- Do you now have a sustainable competitive advantage in the organization?
- Did you build it yourself?
- Are you better off after the transformation than you were before?
- Is the organization more robust?
- Is it more tolerant of variation in the market and your customers?
- Are you more responsive to competitive and disruptive threats?
- Are you more efficient, effective, and consistent in your daily execution?
- Are your customers better off because of your changes?
- Is your supply chain stronger, more robust, and more responsive?
- Has the rate of change reached a self-sustaining level, or will the organization slip backward without your continued investment of CCI?

Woven into the journey of transformation is a natural dichotomy—the gap between intellectually understanding the need for change versus the ability to see the need for change within yourself as others see you. This gap is part of our human nature, and it is evident repeatedly in most improvement actions. A classic example is weight loss—you know there is a need to lose weight, but rationalize the current level is okay and choose not to address corrective diets and exercise habits. This is characterized by individuals seeing the value but not internalizing the need for change within themselves.

Most of us have been part of conversations that focused on a problem, or an opportunity, where the primary cause was always the other organization. It is always easier to point at where others do not perform well, and not to look in the mirror and identify where you are not doing well. This conflict is a divisive one in an organization as it tries to transform itself. It consumes a huge amount of energy as everyone tries to prove everyone else is at fault. It creates conflict and destructive tension in team environments if not squashed at the onset.

One of the best tools, however simple, at the initiation of an improvement action or Kaizen event is the establishment of basic team behaviors such as: egos are left at the door, everything is open for critical review, and identification of contributions to the problem can come from anywhere. There can be no "sacred cows" in an improvement activity.

However, reality says there may be some out of scope issues or environmental conditions that must be addressed. When they exist openly verbalized or exposed and recognized for what they are, move on. It is acceptable to exclude them from consideration, regardless of how strongly the team members pull them back in, and take special efforts to exclude them in the scope, analysis, or root cause actions.

There are numerous examples where everyone agrees on a given course of action, but when it comes to implementation, it always must be done somewhere else first before it can be done at home. A great counter-position to this is that everyone ought to consider their own home turf first before pointing out the shortfalls of others. This means unbiased examination, root-cause analysis, and potential solutioning before expecting others to accept responsibility for their part of the problem.

Work cautiously when using this approach. Frequently, organizations will perform improvement in a microenvironment, and they will optimize the processes, not by work elimination, but by work shifting. Meaning, the work is still present, but somebody else is responsible for doing it now. The problem, or failures, that occurred are now somebody else's failures when an ineffective or inefficient process is moved from one organization to another.

The analogy is like gas in a balloon. If you squeeze a balloon at one end, the amount of gas is moved to the other end of the balloon, which is now bigger and holding the displaced (squeezed) volume. Was the work reduced at one end? Yes. Was the overall work in the system reduced? No. The waste is still present inside the balloon, entropy exists, and the concept of waste removal is absent.

Why is all this important?

It is because human nature seeks to find the easiest way to solve a problem. Likely, that means it belongs somewhere else and it is not your problem. Alternatively, you should not be doing this work because it does not belong to you, and you are going to stop doing it and give it to somebody else to pick up. Neither one is an acceptable answer.

So, how do you know the transformation is taking place? Even the organization has invested a lot of effort into identifying key gaps in building solutions to bridge the gaps for conquering the challenge. In Table 6.1, there are examples of simple evidence—tattletales—that should let you know if it is working, and if it is not, what needs to be adjusted. It breaks the tattletales into green flags, yellow flags, and red flags.

When you see evidence of a green flag tattletale, celebrate and reinforce it with the organization. Focus on how to make the tattletale part of the cultural norm. If you see yellow flags, they should trigger a natural query to find out what is happening and what roadblocks are preventing more rapid transformation. Consider segmenting the information by organization function or site to see if the flags may be uniquely oriented.

Generally, red flags are indicative of wholesale misses in the approach and should require a review of the architecture, its plan, and methods of deployment. Use caution when looking for these flags. In the beginning of the transformation, you will see a large number of red flags. As time passes and maturity occurs, the red flags move to yellow and then green. The speed at which red flags transition from red to yellow and then green is unique to each organization. The rate of change is based on its environment and constraints as well as the crafted rate of change in the transformation roadmap within the plan. It is only then, as the organization matures through the roadmap milestones, that concern escalates when red flags never leave or they only change to yellow and remain that way.

In Table 6.1, many examples of green, yellow, and red flags are offered. A green flag that deserves more exploration centers on the clarification and

Table 6.1 Tattletales of Organization Transformation

Green Flags	*Yellow Flags*	*Red Flags*
• Knowledge sharing and community of practice forums are widely used to communicate learnings and lessons. • Innovation is considered after OTF and simplification is complete. • Process and technology reuse is considered first in the list of potential solutions. • Stealing with pride is recognized and rewarded to reduce resources required for solution. • Lessons learned are the first point of reference whenever a new problem is discovered; then, original research follows as needed. • Rate of lessons learned being captured and shared is constantly increasing. • Problems rarely occur, and never with the same root cause. • Challenges to roles, responsibilities, authorities, and accountabilities are non-confrontational and based on defined criteria.	• Subject matter experts are underutilized. • Innovation rarely surfaces as a solution. • Solution team's recommendations are often second-guessed or overturned by champions with little just cause. • NIH not invented here is typically characteristic MA improvement actions. • New problems frequently have considerable analysis effort expended before referencing lessons learned; read lessons learned are not used as part of new problem analysis. • Rate of lessons learned is declining. • Same problems repeat themselves but may be with a different root cause. • Criteria are formed to establish where and who should be accomplishing targeted work.	• SMEs are almost never included in new solution teams. • Innovation is not part of the disciplined solution adding approach. • Solutions are directed at improvement teams from management with little or no regard for SME inputs or recommendations. • NIH is a cultural norm. • Open resistance to using lessons learned, everything must be reengineered to arrive at a solution. • Rate of lessons learned has all but stopped. Same problem repeats itself with the same root cause. • Finger pointing on who should own or do work is prevalent across the organization and in the supply chain.

division of roles, responsibilities, authorities, and accountabilities for work in the overall value chain. At this point, you should begin seeing evidence of challenges to processes and work tasks regarding duplication, ownership, and optimization. This is a natural escalation during more complex waste elimination activities where the improvement teams have moved from low-hanging fruit, through the great wasteland, and into the tough ground of opportunities.

Examples might include the natural challenges to eliminate redundant checking of input quality for requirements, expectations, or forecasts from the customer base. The likely result of these challenges would identify that it is the customer's responsibility, and they are accountable to assure in-feed correctness. The work to validate correctness and completeness must be accomplished by the customer prior to sending the in-feed to your organization. This moves the work to the proper organization where it belongs, downstream to the customer, and frees your resources to accomplish more value-added tasks that the very same customer is actually buying. Similar examples can be seen in the clear definition of work between your organization and your supply chain partners. There is no doubt that in the early stages of maturity, your organization is guilty of doing the same thing to your suppliers as the customer did to you.

Inverting this concept to incoming deliverables, your organization is likely inspecting and validating incoming deliverables from your supply chain partners. Why? It is the supplier's responsibility to inspect, validate, and assure its deliverables meet your requirements, and the deliverables should require no additional investment of non-value inspection resources from your organization. In truly minimized processes, inspection and validation are non-value added tasks. You are assuring the quality of incoming deliverables from your suppliers by pushing upstream to the suppliers work that appropriately belongs to them. Again, this frees your resources to accomplish work that is more valuable for your customers. This holds true for the relationship between your organization and its customers, too.

In Figure 6.2, the analysis, segmentation, and assignment of tasks appropriately upstream, midstream (your organization), and downstream are tattletale evidence that your organization's transformation is taking hold and beginning to provide significant impact to effectiveness and efficiency. It is important to recognize that while this activity mirrors a waste elimination pitfall called "squeezing gas in the balloon," it actually is a proper activity when conducted with specific intent (versus blame), clarity of scope, and established criteria for decision making regarding

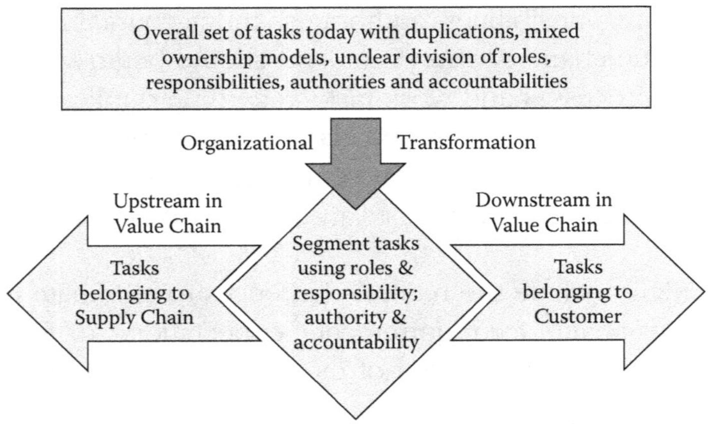

Figure 6.2 Transformation of value chain accountabilities.

roles, responsibilities, authorities, and accountabilities. The division of work outside your organization, and some back in, provides more complete control and opportunities for improvement across the entire value chain ecosystem.

Be leery of overlapping job titles and responsibilities across the entire supply chain. Use the criteria to appropriately determine where the work is performed, and what aspects of time, cost, and quality measures are used to evaluate the performance at this step, and then ensure that there is a defined owner for the results.

Many organizations have outsourced direct labor without evolving their process for managing the work. If people inside your organization think that they are responsible for the quality of another team outside of the organization, then a problem exists with roles and responsibilities. People doing the work must understand what acceptable quality, cost, and on-time delivery (OTD) look like. You cannot inspect quality into the deliverable from your office, which is often thousands of miles away. Figure 6.3 refreshes five significant areas of learnings acquired within a Learning Organization that need to be migrated into the supply chain, phased so they make sense, and will be of value for the supply partners.

The same is true when you look up the supply chain into your customers. The minute that you start to emulate the VoC's performance in all aspects of the organization that interfaces with them, you start the journey of simplification upstream into the customer's environment.

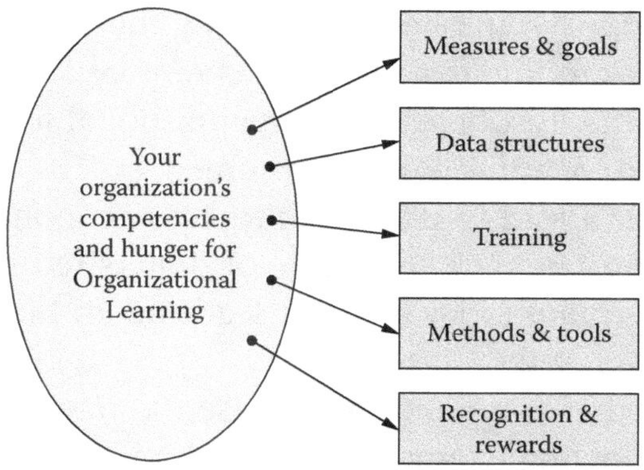

Figure 6.3 Migrating learnings into supply chain.

6.2 Accelerate the Applied Rate of Learning

By now, the organization is using the new tools to find and remove waste, streamline processes, and increase overall efficiency, effectiveness, and consistency. This sounds good, but it is not enough.

When asked about accelerating the rate of learning, a few questions quickly emerge:

- How is this done?
- Why is the rate of change now not good enough?
- How much more is it going to cost?
- How much more can the organization absorb?

These questions, and many more, provide a baseline to examine any new changes within the organization, the rate at which they occurred, and how much more it can absorb.

Accelerating organizational learning is composed of measuring the current rate of learning, determining the desired rate of learning, and developing a solution to change the rate of learning between current and desired. You need to recognize that to change the rate of learning requires time and resource investment. There is a rate of return for each investment and it must exceed the investment cost; otherwise, the improvement or the change is only being accomplished because it can be.

Throughout this book, there has been discussion about the dichotomy between "should be" and "what you are capable of." This is one more example. There is a saying, "gilding the lily," meaning "make it better is the enemy of good enough." It implies that making the value creation process better without understanding exactly what is required for the intended purpose is a waste of time, resources, and organizational focus because there is no need, expectation, or value by the customer.

You can no longer afford to do that in today's economic times. Every dollar spent must translate into a multiple of dollars returned. Be critical and practical as you determine today's rate of learning and its efficiency for the organization and its growth.

If you determine that you need to increase the rate of learning within an organization, the following examples provide hints that have been used to accelerate the rate of learning:

- internal certifications of proficiency
- internal competitive process at organization unit, functional, team, and individual levels
- active benchmarking from organization to organization
- entrance into competitive awards
- internal recognition processes based on rate of return as validated by finance
- special recognition or treatment for Lean champions and Lean leaders (e.g., special badges, pens, titles, anything else that identifies their unique and special skills)
- dedicated management attention to details
- leadership and management demonstration of practices that are expected throughout the organization
- cohesive deployment without exception across the organization
- proper tailoring of methods, tools, and training based on work type

6.3 Leverage Learning with Lean into the Supply Chain

How do you break the traditional model? The first step is to accept that the extended group of downstream suppliers is responding to your direction.

- What have you told them to do?
- Are your actions and words aligned?
- Is your best performing supplier getting more or less of your organization?
- Is it a surprise to them when they win or lose a new opportunity?
- How have you communicated scheduled priorities?

- Are the quality requirements that are important to your end customer clear and simply communicated?
- Are your cost targets predicated on sharing savings from reduction in waste or based on trying to erode your supplier's margins?

The point being highlighted is you need to see your organization through the eyes of your suppliers to determine the opportunities for improvement. Let us assume that you have established a long-term relationship with your principal suppliers. How do you get them to buy into the LWL approach? Get ready to spend some more of your CCI!

> Suppliers and especially manufacturers have market power because they have information about a product or a service that the customer does not and cannot have, and does not need if he can trust the brand. This explains the profitability of brands.
>
> **Peter F. Drucker**

You should have tangible improvements inside your organization to share with the suppliers. The evidence of these successes will be visible to your supply base. You face the same dilemma with deploying LWL with your suppliers that you had when you initially started within your organization.

- Have you clarified what the journey means to them?
- Do they have the resources to start the investment in the people who are doing the work?
- Have you demonstrated your commitment to their financial well-being?

Remember, they have seen numerous approaches from customers to help them "share" in savings that simply resulted in reduced profitability margins for the supplier.

At the end of the day, your conviction that LWL will reduce costs for both organizations will be required to get them to start to move. Remember, "If it doesn't jingle, it doesn't count."

Let us examine your supplier's environment. They probably serve a number of customers. The criteria for success with each of their customers are different, each with a performance measurement system that probably varies greatly. The suppliers probably have direct and indirect resources that are shared across multiple customers, so their standard work will be defined in very different terms than yours. The way they reward and recognize personnel is not the same as the process used in your organization.

Where to begin? As you articulate the direction you want them to go by LWL, you need to show how this is going to be a journey, where both parties are held accountable to each other for progress. The next item is to enable them to "pocket," or keep, some of the savings from their first successes so that they have a way of funding further developments. In the simplest form, your commitment from your team to them could be a 50% reduction in touches in a process, defects in documents, or instructions that your team issues, or simplification in the flow of data.

You will need to drive these improvements through your internal team, but what is it that you are asking them to do? Deploy LWL!

They can only do this by driving the tools and capabilities to the people doing the work. What does this look like as it unfolds?

- They end up with a workforce that works on your deliverables, that acts and improves processes different from their peers.
- They have been well trained to identify waste and detect defects in your products and processes.
- As they eliminate some of the waste, there may be fewer of them.
- However, because they are better trained and can perform multiple functions (cross-trained), they are more valuable to the supplier and better compensated.

The supplier has now evolved into a team that works differently than their peers, they are compensated better, have much more say in how their jobs are done, and have a happy customer—this sounds like the place where people want to work.

With this view of the possibilities, there are so many ways to determine if you are making progress.

- Have on-time delivery, deliverable quality, and costs continued to improve?
- Have the revenue per employee, space utilization, and asset utilization improved for the supplier?
- Has employee turnover reduced?

The implications of working with your supplier this way are far reaching in your own team. Today, many staff members may assume a dominant role in communicating with suppliers. In some cases, the perception of power comes from the belief that they have more knowledge and subject matter expertise, or aptitude, as it relates to the specific discipline with which they are dealing, in the general case or as it relates to your deliverables.

This antagonistic approach to managing the supplier does not fit in the LWL philosophy. Staff members must evolve their roles, and coach or mentor the supply base. The approach to achieve desired outcomes must be collaborative, with clear definition of roles and responsibilities and well-defined quality metrics for all deliveries in both directions. The relationship must be synergistic, not parasitic. Here again the organization will have to demonstrate maturity to let go of some of the responsibilities that are historically held, entrusting the supplier to deliver the desired results. The individual contributor's role becomes one of oversight and process evolution focused on the internal deliverables from your organization to the supplier.

Your organization's evolution of responsibilities can be characterized as:

- Migrate from inspection and checking to process oversight and monitoring the rate of improvement.
- Externally and internally, your team should be moving from doing and checking the work to improving the quality of what they deliver to their suppliers.
- Validate the tools and process capabilities that are being used to evolve the processes of the organization.

One of the principal changes in expectations of the supply chain is that they must move from measuring percentage performance (e.g., percentage failure that provides little discrimination on what to fix) to measuring opportunities for improvement (defective parts per million [DPPM]) (high degree of discrimination on what to fix and data analysis is the map to determine what gets fixed first).

When setting goals in an ambiguous situation, always opt for the most simple:

- If performance is above expectations, strive for "twice the good" (double the yield while maintaining quality and cost).
- If performance is unacceptable, declare "half the bad" (cut defects in half each measurement period).

Measure your supply chain as you measure yourself. A key concept is driving expectations into the suppliers that they measure and control their process performance, and that you only monitor their measures as an oversight as opposed to inspecting incoming deliverables for conformance.

A tattletale of the transformation moving upstream into the supply chain is when suppliers migrate from measuring the quality of their process using capability measures (process capability) and deliverables using percentage defective to measuring DPPM.

Why is this tattletale relevant? Noted in Figure 5.2, the focus of Wave 2 is process control and capability management. Their maturing measurement and control systems show evidence that process control is taking precedence over process execution.

6.4 Standard Work

> Where there is no standard, there can be no improvement. For these reasons, standards are the basis for both maintenance and improvement.
>
> **Masaaki Imai**

What is standard work? The term is used frequently, yet it lends itself to wide interpretation and is somewhat hard to pin down. In fact, it should form a cornerstone in the way that you do your work.

A brief definition of standard work is:

> Normal tasks done with the least amount of waste possible; it focuses on the work done by the workers, not the equipment or materials.

From the definition, standard work can take on an infinite number of forms. Why is standard work important?

- It reduces variability.
- It increases control.
- It reinforces discipline and execution rigor.
- It enables consistent management across the organization.

Standard work can be categorized into two basic forms—worker standard work and management standard work:

- **Worker Standard Work (WSW)**—The attributes of WSW include repeatable processes or services, high frequency execution sufficient to establish meaningful performance metrics, processes capable of measuring variation in capability, and consistent work accomplished across teams, work units, or divisions of an organization.
- **Management Standard Work (MSW)**—MSW is modeled after WSW, with the exception of its focus. The attributes of MSW include daily, weekly, or periodic management review practices, people management processes, communications—up, side, and down—in the organization, and general exception management work (recognizing that the content is different for each, but the process is the same).

Briefly, a few of the benefits of standard work include:

- common process, collateral, and deliverables
- common language eases communication missteps
- improving one process affects the shared processes across an entire organization
- reduced variation in execution and outputs
- more consistent reporting and data analysis opportunities
- reduced overhead to manage just one process

One of the cornerstones of being an effective and efficient organization is the use of standard work—both WSW and MSW. Both are centered on the concept of consistent practices executed in consistent ways and monitored and controlled using consistent methods.

At the worker level, this is the most common area for consistent practices. They cut across all functions within an organization and are typically those processes used in the value creation process for deliverables to a customer's needs. There are ancillary processes that are required to enable and support value creation processes. Examples of these are human asset management, financial support and services, facilities management and maintenance, security, and a host of others. Each of these processes, while some can be done different in terms of content, is still executed in the same way. It is at that level that standard work is categorized, documented, and monitored. An example is exception management.

At the management level, there are standard practices that everyone in management leadership roles performs. They range from allocation of resources, financial management project identification and oversight, developing strategies to meet the organizational mission, and a wide range of other tasks.

Let us look at a common one that all managers deal with—exception management. By the nature of and using the escalation processes organizations have in place, all members of management from supervisory level up receive escalations that turn into exception management. It is easy to see that the content of each one is likely to be different, but the methods, the systems, and the control points are likely common and shared across the members of the management and leadership team. These processes can be documented and controlled. Through the application of MSW, you should be able to identify which items are truly outside of the process control and then determine if the improvement plans already identified address these items. Finding an exception that is not covered then becomes a gem for the organization to investigate and determine how and where the corrective actions fit in the hierarchy of the improvement plans.

Another example is periodic performance reporting within their organization and its aggregation up the management chain to higher levels of leadership. While the content is different for each organization (in terms of measurements), the management and control techniques follow similar methods. Those methods can be standardized and routinely accomplished across the spectrum of the management team.

Exhibit 6.1 is a representation of a simple project report call an "A3" (drawing from the paper size used in its format) used as a consistent approach to manage periodic project reporting.

A positive attribute of using standard work is that it is formulaic in nature, and drives consistent and standard practices on development and implementation at all levels. Tying back to EECO-feedback organization (i.e., effective, efficient, and consistent organization) using this approach augments your drive for exceptional performance.

6.5 Clustering Thousands of Small Fires into a Forest Fire

> The image is more than an idea. It is a vortex or cluster of fused ideas and is endowed with energy.
>
> **Ezra Pound**

Report – A3 Project Record Dept: Title:

Team: Status: Date:

Background, or Reason for Change:	Improve: Show or Describe Counter measure(s) and Final Solution
Define the Problem:	
Measure: Go See – Show or Describe the Current State	
Analyze: Identify the Root Cause(s) of the Problem	Control: Standardize Improvement and Adherence to Process
	Follow-up:

Exhibit 6.1 Improvement report (example).

There is a desired end game description that characterizes transformation in organizations as a landscape with thousands of brush fires (improvement actions). Eventually they merge into large blazes. That act of merging is an event horizon, where suddenly the use of Lean as a learning tool is rampant. Let us not characterize it as out of control, let us characterize it as self-initiating, spontaneous, incendiary, and feeding a deep need within the organization to fix its constant dissatisfaction with the status quo.

Clustering is linking the many small fires (brush fires) into the larger forest fire. The key concept is about identifying and developing synergy between small improvements to solve the larger opportunity. This is about the approach of converting many small, unconnected improvement actions into a larger, cohesive tide of changes that build momentum and change the course of the organization's culture.

In the beginning, it takes a lot of energy to start the small fires, first one, then another, and so on. It takes a network of aggressive Lean leaders to initiate these improvement actions.

So, how do you get it from being a set of independent and non-synergistic improvement activities into an aligned, cohesive, and spontaneous change-hungry improvement behavior across the organization? A wide range of approaches has the ability to help move the transformation closer to the desired state:

- Lean learning forums
- Learning and collaboration exercises
- Lean leaders embedded in most organizations
- Communities of practice
- Embedding expectations for LWL in annual performance reviews
- Improvement team competitive events
- Recognition and reward for exceptional improvement results

What do thousands of improvement fires look like? The easiest way to describe it is that when you open the door to your office you see evidence everywhere you look. You hear discussions in most meetings that have elements of improvement and waste reduction. When you walk the floor where the work is done, you see people and teams focused on removing the last element of waste in the work they do.

Figure 1.1 shows the evolution of improvement efforts from independent and isolated in nature (on the left) to synergistic, aligned, and coordinated (on the right) for superior results. By using many of these methods, the migration from left to right is accelerated and embedded in operational behaviors.

6.6 Reinforce Good Decision Making and Risk Taking

> Failure is the opportunity to begin again, more intelligently.
>
> **Henry Ford**

Good decision making and learning must be appropriately rewarded. There must be no punishment for decisions that did not work, but were done using good data and critical analysis.

LWL does not exist in a vacuum. Most of the good management practices and techniques you have learned before apply. Reinforcing the behavior you are looking for is critical to the adoption of LWL. You can only do this by encouraging people to try and if they fail, encourage them to try again.

Try-storming is, by design, an iterative process and a key tool in the process improvement practices found in LWL. If failure has any negative connotation attached to it, people will not adopt and use this proven method to implement change. When you think about sharing results, encourage the team not only to identify those areas that were successful, but also reinforce those examples where solutions did not produce the desired results and required rethinking to achieve the desired outcome.

Having the courage to try and fail is a key tattletale of the organization's maturity. Many of your team, especially management, may have basked in perceived omnipotence. That does not work in LWL. Failure is a critical part of learning.

The better the teams are at sharing best practices, including the insights gained from failed endeavors, the more representative they are of a mature organization's capabilities. As the leader of the group, you are allocating the resources of the organization.

At this stage of the journey, you should have moved from measuring metrics to monitoring processes that have well-defined control limits and capability measures. This implies that there is a well-defined path for any and all exceptions. The team will also be flirting with establishing a competitive advantage on each dimension of performance. Again, you are not in a quest for perfection but rather the ability to provide value-added deliverables to the market that assures a sustainable competitive advantage. Once this level of performance is achieved in one area, more energy can be injected into other areas or processes to get them to this level. The assessment and prioritization of focus should be a natural replacement for the void left by the elimination of many unproductive, historical managerial functions.

Your ability to recognize that you are the industry leader in performance, along with monitoring this level of performance, needs to become a core competency. The rate of evolution in your competitors' ability, as valued by the customer, will help guide you through the level of investment in improvements required to keep your advantage. Failure to recognize when you are "good enough" leads to the risk of introducing non-value added waste into the solution.

Since data is the source of the evaluations of process improvements, the scorecard of improvement is well defined before beginning the process. The way you can help convince the team that failure is an accepted part of this process is to spend as much time reviewing the setbacks in the concepts that did not work as you did describing shared learnings across the group.

6.7 Benchmark and Compare for Higher Performance

> All successful companies are constantly benchmarking their competition. They have to know what they have to match up with day-in and day-out if their company is going to be successful.
>
> **James Dunn**

In the beginning of this book, the stage was set that your organization is, most likely, a good performing organization. This means that it is generally meeting expectations of stakeholders, customers, and leadership. Assumptions include key measures are in place, feedback systems exist, and some preventative methods are present to identify root cause and solution for permanent fix.

What is missing from this picture?

The ability to improve progressively, in an already good performing organization, frequently requires a view outside the organization. Why? It is human nature that when you design, build, and execute processes and systems, you become so close to the way it is supposed to work that you are blinded to better alternatives. Other approaches to solutions may be more effective, more cost efficient, or easier to implement. How would you know if you are blinded by personal bias?

Here is where benchmarking comes in to provide comparative information, methods, and systems. Benchmarking is not a new term for most, but requires a little time developing the concept based on the premise of the book, and its value in accelerating the learning capability and building a sustainable competitive advantage.

Benchmarking, while often used interchangeably, is composed of two elements: the action component—benchmarking, and the measurement component—the benchmark.

Briefly, for the purposes of this book, both will be defined as:

> Benchmarking is the process of partnering with one, or more, organizations to find the best methods, systems, and practices among them, and then identifying the gaps between your current approach and the best practices.
>
> Benchmark is an actual measure of performance (an artifact of execution) for a given product, process, or service.

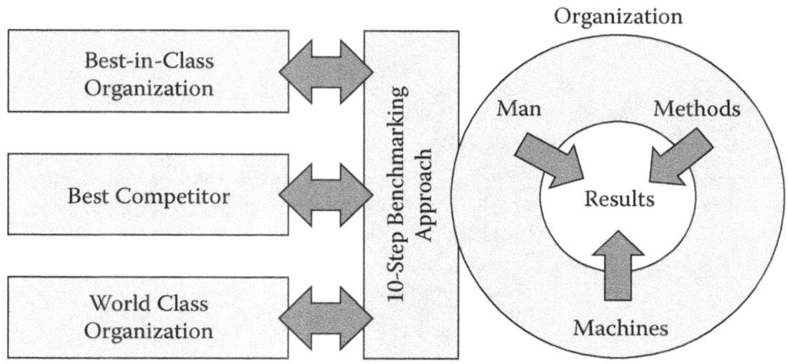

Figure 6.4 Benchmarking requires external perspective.

Both elements are equally important and necessary to truly understand what the best methods and results are for any approach that is being evaluated for any area.

Benchmarking (action component) relates to the process, the 8-step process used (author equal camp) to define the subject, partners, discovery methods, and results that are capable of guiding changes to methods, alternative solutions in products, or evolutionary changes to services delivered to customers. In Figure 6.4, the fundamental structure of external benchmarking is composed of the organization, its 8-step benchmarking approach, and the external organizations used for learning.

The benchmarking approach is capable of examining any kind of workflow, product, or service. In Figure 6.5, the 10-step benchmarking model is adopted from Robert C. Camp's book, *Benchmarking: The Search for Industry Best Practices that Lead to Superior Performance* (Camp, 1989). It follows a logical flow through each step with defined inputs, outputs, and approvals required. Typically, there are standardized forms for data collection, partner identification, and agreements for nondisclosure, analysis, reporting, and sharing of findings.

A word about scaling the benchmarking process: If you use the full 10 steps on a detailed process, it will consume time and resources commensurate with the scale of benchmarking scope. Early on, use the full process sparingly; use it only on core processes fundamental to the organization's mission success. Later, as the core processes and systems become well tuned, the approach can be used for incrementally lower processes and methods.

Taken to extreme scale reduction, the process of benchmarking can be minimized from full strength (all 10 steps) down to something called industrial tourism.

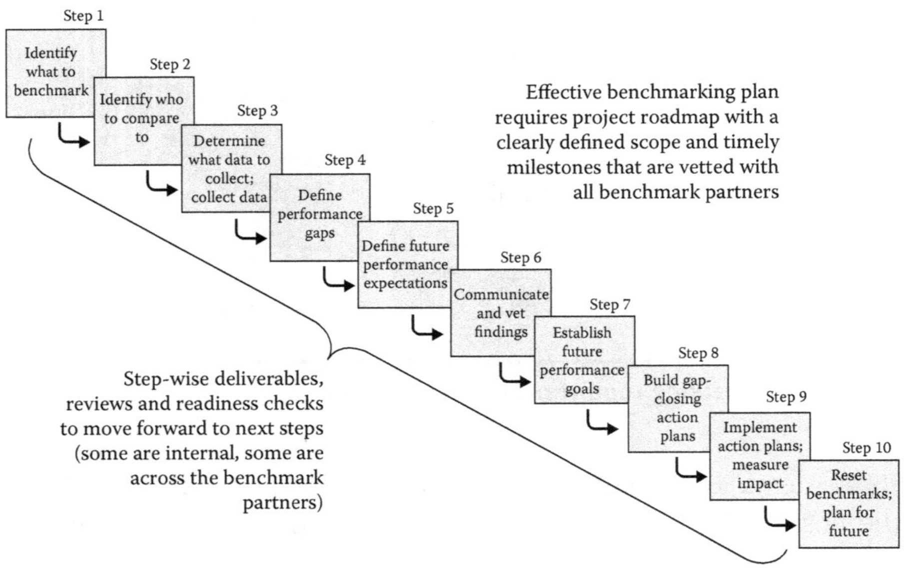

Figure 6.5 Benchmarking process steps.

What is industrial tourism? In today's time with Internet search capabilities, content-rich resource organizations that sell benchmarking services, and the growing presence of government and educational institutions, knowledge becoming publicly available you can now gather a tremendous amount of information very quickly in a short period of time. Assuming the use of good filtering methods to exclude unsubstantiated work, you can generally find a large portion of information needed without expending a lot of resource. With the economy getting tighter and resource loads getting heavier with available bandwidth for improvement being squeezed, frequently industrial tourism offers a viable and totally suitable solution to benchmarking needs.

Benchmarks (measures component) are the findings, or results, of the process of benchmarking. They are quantitative, or qualitative, in nature, taking the form of a process parametric, product attribute, or service attribute, plus process and system methodology descriptions. These are only abbreviated classes of examples because the range of measures and results is limitless as you look across the breadth of organizations and the landscapes in which they operate.

As mentioned in the benchmarking process, frequently benchmarks can be obtained through industrial tourism. Affiliations with various associations, clearinghouses, and standards bodies are good sources of benchmark data.

Benchmarks are the deliverables used for process and system improvement. As noted in the benchmarking process, frequently benchmarks from outside your industry present the greatest opportunities to challenge the status quo. Examples of where this is true abound across the range of organizations. Consider the learnings, process descriptions, and methods from automotive racing teams that achieve under 7-second pit stops. They are adapted and used by some airlines to streamline and improve the turn time an airplane makes from its arrival, unloading, loading and servicing, and ending with departure

Benchmarks should be used to frequently challenge the status quo of work teams, enabling them to achieve phenomenal step functions in the way the work is done. While evolutionary changes are normal and accepted, step functions are the Holy Grail and benchmarks help set the bar for improvement results.

When similar tasks, regardless of industry, are compared and one is done at a rate 10 times better, 10 times faster, or 10 times cheaper, questions are raised and the challenge is opened. Most organizations are highly competitive by nature, and can use this cultural norm to leverage the competitiveness, feed the winning spirit, and use benchmarks to set new and challenging aggressive goals. You and your leadership team are obliged to give teams the free reign, authority, and accountability to investigate, improve, validate, and lock in process changes to meet, or exceed, the aggressive goals founded on benchmarks.

Previously, try-storming was explored, but this is where benchmarks and try-storming become integral parties to step-wise improvement actions. Benchmarks establish a significantly higher goal or level of execution performance. An effective learning organization utilizes the findings, information, and knowledge gained through benchmarking to feed fast prototyping processes, or methods similar to try-storming, to accelerate the rate of learning.

6.8 Reinvent Work and Job Skills

> You have to reinvent yourself every day, and that means being a pioneer.
>
> **Frederic de Narp**

In the Learning Phase, what does all of this mean to the average employee? Based on what you have read thus far and are now implementing, there are two points to remember:

- changing the way the work is done through process streamlining with OTF, and
- mastery of skills required to analyze, change, and perform the value-added work that remains at elevated levels of performance.

Doing the work is different; the very nature of using Lean to find waste and remove it, and then to optimize performance means that the work being done today to achieve the same output has a much higher percentage of value-added work, with a greatly reduced amount of non-value added work (waste). Fewer steps equals higher quality, faster cycle time, less cost, and less resources required to complete the task.

Mastery of new skills is the one that most directly impacts employees; it is the one they care about. Up-skilling partially addresses one of their first fears at the very onset of the transformation process. As part of the transformation plan, you have to be able to describe and demonstrate that fewer resources required do not mean a reduction in personnel. It means a reduction in non-value work, and that opens up current personnel to learning and using new skills for a higher rate of return to the organization.

Let us address the more kinds of work first; this means that for any given task, less resource is required. With less resource required, this opens up every individual to do new or additional value-added tasks. This is a natural feeder into the reinvention of skills (up-skilling). It implies that many workers will migrate from being dedicated to a single task and having deep knowledge to being a team member who does a wider variety of value-added tasks well using their deep existing knowledge plus augmentation with newly acquired skills. As leader your job is to reinforce the work that the organization must keep doing and identify the work it can no longer afford to do because the overburden of waste is diminishing the organization's overall performance. By using the LWL approach, the organization can quickly optimize existing tasks and add new tasks to create a more capable and robust organization.

This means new skills are needed throughout the organization, but it does not mean new skills are needed at the same time. As leader, one of the important points to convey to your workforce is that the organization is dedicated to investing in their personal growth, increasing their personal value to both the organization and to themselves, and setting expectations that the skills will be best used by the organization and its customers.

Referring back to Phase 2, Architecture and Planning, one of the critical elements was a well-designed deployment approach. In that discussion, training the core philosophies, principles, and practices of LWL across the entire organization was explored.

It is important to recognize that as you go through the five phases, you are going to do them repeatedly. You have completed the first loop; now it is time to get ready to go around again. What this means is that as you have optimized processes, removed wasteful work, and opened new opportunities for skills, you have adjusted training plans for the next round. This implies that new or improved processes have been identified, and in order to execute those new tasks or processes, it may require new skills for the existing workforce. It may also require moving employees around in the organization to best leverage their backgrounds and existing skills, some of which may not be used well today.

Human nature being what it is, employees in the organization today do their jobs typically because of their knowledge, comfort, role, and responsibilities. LWL is going to push them to the brink of being very uncomfortable.

Why? Because the organization is going to ask them to keep doing today's tasks but at higher levels of performance, and learn how to do new value-added tasks as additional responsibilities.

As leader, you need to offset fears with opportunities, offset apprehension with excitement, and offset lethargy with challenges. Part of your communication plan needs to include what new skills may mean for them, how they may be able to leverage them for internal opportunities and advancement, and how their personal worth is being elevated through the organization's investment in them. Consider including topics such as:

- Increasing the rate of reinvention at the individual level.
- Reaping the benefits for the customer.
- Reinforcing that transformation is everyone's job, not just Lean leaders.
- Workplace reinvention:
  - Answers how will you leverage my free time to benefit the organization and myself?
  - Does not mean losing your job, it means doing higher value-add work and learning new skills.

- Sets higher expectations of the workforce to reinvent themselves enabling them to exceed new organization models and changing performance challenges.
- Assures constant evolution to work processes and deliverables.

6.9 Measure and Share Progress

> When it is obvious that the goals cannot be reached, don't adjust the goals, adjust the action steps.
>
> **Confucius**

A principal expectation in the transformation journey centers on measurement visibility. Noted before, a good performing, learning organization typically keeps its core process performance measures visible for its workforce. The expectations are that through visibility, the workforce will react and respond with changes in the way they are working. When variations exist, analyses are performed to determine adjustments required. Where process escapes occur, they can analyze and identify root cause so a sustainable corrective action can be put into place.

There is a simple analysis framework used in a widely adopted Performance Excellence Framework developed by the National Institute of Standards and Technology (part of U.S. Department of Commerce) and used to evaluate awardees for the United States government's highest quality award—Malcolm Baldrige National Quality Award (Baldrige Performance Excellence Program, 2011).

Contained within the performance excellence criteria, an evaluation model is used to gauge an organization's performance measures and results, and it is made up of four components:

- **Level**—measures of performance, or execution, of a specific system, process, or deliverable.
- **Trends**—repeated measures (levels) over time.
- **Comparisons**—information, benchmarks, or competitive intelligence for a specific system, process, or deliverables gained from a source outside the host organization.

- **Integration**—measures address important customer, product, market, process, and action plan performance to requirements important to the organization's mission.

Let us amplify each of these four components to understand what they look for, how they are used, and the cross-linkages between them.

Level—Is the most common component, and essential for the other three components to be present. Levels are traditionally measures using current performance for the latest measurement. Unfortunately, for most levels measured in organizations they use simple counts or averages. They are traditionally not normalized or weighted. When using only averages, considerable information is hidden from view; variation contained within the process, both common and special cause, cannot be seen. Without the understanding of variation, it is hard to detect in or out of control processes. It is impossible to characterize the capability of a process without understanding its natural variation, its distribution, and process expectations or limits. All of these are contained in the data stream used to measure a level.

Trend—While simple to understand, trends are often argued at length. For the purpose of this book, trends are considered to be characterized when the number of data points is five times the measurement frequency (meaning five or more data points). There are statistical camps declaring 8 data points, 10 data points, a year's worth of data, and others, all of which have elements of validity, but there is no fixed rule applicable across all organizations. What is important for your organization, when it comes to measuring trends, is to present enough data to determine shifts, aberrations, or cycles of performance. Trends are frequently analyzed using linear and nonlinear characterization techniques. History shows that first-order analysis of data points is normally adequate. Again, the criticality of measures is unique and tailored to each organization. The essential rule is if you are measuring something and the measure changes and you do nothing, you are measuring the wrong thing.

Comparison—Provides the foundation for whether the goals for an indicator have merit or are aggressive enough to drive higher levels of performance. Frequently, comparative key measures are derived from benchmarking or shared benchmarks. Sometimes comparative measures are derived through personal networks, industry forums, or public meetings.

Integration—Identifies measures that are central to mission success, are valid aids in future performance, and are harmonized across processes, divisions, work units, and geographies in support of organization-wide objectives and goals.

In all cases, measures should be visible to the organization, depending on what is being measured, where it is being measured, and the audience. Measures derived from work units or work teams must be visible to those teams at a frequency high enough that enables reaction and response. Examples of visible measures include:

- totem poles
- common add-on boards
- posted dashboards
- process performance charts
- virtual performance portals housing process-specific, detailed metrics
- shop floor controls visible to everyone

As noted before, data is the backbone of effective decision making. Measures are the way we communicate those data. There are many examples of highly effective work teams that use such techniques as standup meetings daily around performance measures, using the effort to focus on throughput, quality, repetitive issues or problems, root cause analysis, and resolution actions.

6.10 Communicate and Walk the Talk

> Precision of communication is important, more important than ever, in our era of hair trigger balances, when a false or misunderstood word may create as much disaster as a sudden thoughtless act.
>
> **James Thurber**

Remember the credo for the LWL journey:

- Raise the bar on your team—no more money, no more people, no more space.

- Celebrate every win; the journey is tough.
- Face new challenges with self-confidence.
- You are never good enough for your customers.
- Continuously successful enterprises are perpetuated by continuously unsatisfied leaders.

Make sure that your MSW is current and used with your staff. Again, this is a public display of what you are focused on and where you think the next improvements will come from. This alone will send a strong message to the team of your conviction to LWL.

Participate in the focus group activities. Drop in on the cross-functional standup meetings. Use the information from the tools of visual management to recognize achievements for the people doing the work. Driving a clear consistent message of LWL remains critical. By engaging with the team on multiple levels, helping assess and evaluate the nature and quality of the material used to help train the teams will keep you grounded on where the challenges are and what the capabilities the team already possesses are netting in the way of improvements. Be aware that the multitude of improvement projects going on will easily dwarf your ability to review all of them.

Use an open mind of discovery so you can sit with teams to find out what they are working on. The MSW will outline the big elements of work, but not the thousands of little improvements that are being made. Here you need to help encourage the groups, and focus them on sharing great ideas. Many improvements are directly applicable to other parts of the organization, other sites, and other business units. By leveraging these learnings you will compound the rate of development for the entire group.

By now, you should have a good idea where any additional changes in management will be required. Not everyone has to be a Lean SME, but they do have to demonstrate skills when managing projects (e.g., focusing the efforts of these cross-functional teams). While it is interesting to note that the management team will lose personal cash from reduction in relevance of their SME (in many cases, the team will outpace their managers' understanding of how things are done), their general organization acumen will become more valuable in focusing the efforts of the group. In other words, the skills that were valued in previous promotion decisions will have to be reassessed as you look at the new organization paradigm.

Key Messages in This Chapter

1. Results are starting to lead to differentiated performance, leading to a sustainable competitive advantage.
2. The emerging challenge is to propagate the learning curve on multiple fronts of the organization, and up-stream and down-stream to suppliers and customers alike.
3. It is time to reevaluate, refresh, and reinvest. The cycle of your journey is a continuous one.
4. Try-storming is an essential tool that promotes rapid identification and solutioning, delivers increased results for the effort expended, and helps the organization learn at ever-faster rates.
5. Tattletales are simple, readily evident changes in behaviors that give you the pulse of the organization's rate of learning and adoption of Lean practices.
6. Not all work being done today should be done by your organization, segment the work critically, and find out who should really being doing the work; migrate appropriate work upstream and downstream to your supply chain and customers.
7. Benchmarking is critical to being able to compare current performance versus your competitors, or against the market. It requires focus and energy to be impactful—use the approach when necessary to set the right meaningful goals.
8. Reinvention of work and job skills is required to help keep organizational fear in check. Approach streamlining openly with the organization and define how liberated resources will be refocused/reused.

Challenge Actions

This section has been about leveraging the Learnings. It is about recognizing the yellow and red flags when energy and focus on becoming a Learning Organization and execution of Lean practices is waning. Communication and measurement of progress are cornerstones of this section as well as keeping your foot on the accelerator to increase the rate of learning and improvement.

Choose actions totaling at least 50 points, ones that focus on increasing the organization's awareness to see traits and behaviors of organizational willingness for learning and improvement. They should also include actions that help reenergize and increase the rate of learning, communication, and measurement of progress.

Challenge Actions—Leverage the Learnings	*Points*
1. Find evidence of cross-functional participation in objective setting and results delivery. Capture the findings and have a discussion at your next staff meeting, focusing on how to make the practices systemic.	20
2. Identify two functions where layers of management have been flattened due to Lean process minimization and work/task elimination. Determine the change in performance of the function and the trends of projected performance gains. Highlight findings at staff meetings.	15
3. Construct a short list (5 to 10 items) of the organization's transferrable skills.	10
4. Develop a punch list of the organization's transferrable messages that can be used anywhere, at any time. These fit the description of the "one floor elevator speeches."	15
5. Using Table 6.1, add additional tattletales in the green, yellow, and red flag categories that are unique to your organization's culture, operating model and customer environment. Use these in periodic assessments to see if organizational learning is taking place.	15
6. Pick a key deliverable, inventory the work being done by your organization today, analyze each work action to see if it should truly be yours, or it should be done by either your supply chain or your customers. Frequently, work gets absorbed over time that isn't/shouldn't be yours. Segment work by accountable owners into supply chain, your organizations and customers. Develop a migration plan for the supply chain and customer owned actions and begin execution within 90 days.	30
7. Can you identify where reinvention of work or job skills has occurred due to application of Lean concepts and process streamlining? Build a short list of examples, conduct learning sessions with the relevant workers and managers, consolidate the learning, find key messages and conduct management and staff "sharing sessions" to promote work and job skills reinvention.	20

8. Are the performance measures used to manage the organization systemically shared and used to drive improvement actions across the organization? Evaluate what is measured, how the measures develop understanding of levels, trends, comparisons and learning. Develop a short list (3 to 5 actions) that will address measurement shortfalls and implement one each month. Refresh the list when only one item remains.	15

Chapter 7

Wrapping It Together

Change is the law of life. And those who look only to the past or present are certain to miss the future.

John F. Kennedy, 1917–1963
35th President of the United States of America

By now, you have read the recipe for implementing the strategy of LWL. You have assessed whether you and your team have the capacity and the inclination to leverage the strategy. You developed plans, prepared to implement those plans, and started the organization moving down a path of continuous improvement and LWL. You extended your reach beyond the original targeted group on which you focused your efforts, and are now looking across organizational enterprise-wide endeavors. With that as a backdrop, let us review the types of changes you should be able to see in the areas of people, methods, and machines across the organization.

7.1 People

One of the core tenets of this book is that to create a learning organization you have to invest in the people. This fundamental, ground-up approach lends to the uniqueness of the journey. It also adds to the difficulty of execution, and these two factors combine to ensure a competitive advantage because the results are not easily replicated by others. We said repeatedly that the journey was going to be tough, and some would have a hard time adjusting to new concepts. Beyond that, there are a number of other attributes to the new organization that are worth considering as you look to assess the current rate of deployment.

Have you achieved the critical mass necessary for sustainable learning within the group? There are a number of ways to assess this. One of the easiest ways to assess this is to determine how much of your CCI is still required to keep the process evolving.

- Are you being surprised by the number of projects underway in the rate of improvement delivered independent of your involvement?
- Is the cross-learning between sites, functions, and elements of the organization happening at a rate that surprises you?
- Has your primary focus on communication shifted from encouraging people to try this solution to harnessing the capacity and to the strategic advantage that you have developed?

These are some simple questions that help you understand the level of maturity that the organization has reached.

One of the fundamental issues that you will have addressed is the notion of growth and improved efficiency. Very few organizations have a four- or five-fold opportunity for growth in front of them in the near-term planning horizon. Yet, this is the type of improvement you are expecting to deliver with the strategy. This is the fundamental dilemma with which you will have to deal. At some point in time, your improved efficiencies through the reduction of waste will outstrip the growth of your revenues or the capacity required.

If this is a short-term situation, the benefits you have delivered to the bottom line will allow you to absorb these extra employees and continue the journey. If this looks to be a prolonged situation, you are going to be faced with the difficult task of reducing the workforce. In this lies the quandary that has people driving improvements only to work themselves out of a job.

As a leader, are you prepared to handle this situation? Unfortunately, the reality of modern employment is that we should all recognize it as being tentative. As a leader, you are trying to deliver to your team the advantages of learning and growing on a daily basis within the roles they have. If you can be comfortable with the fact that your motivation is investing in the development of each individual, subsequently making them more employable and more valuable to the next organization they work with as well as your organization today, then you can face the dilemma. This not only affects the people doing the value-add work, but also this affects the management team. Be prepared to develop people who will move on to other roles outside of the organization. This will be a testament to the quality of the solution you are developing

By now you should have seen a considerable flattening of the organization. The traditional hierarchy frequently in place has been replaced by a matrix structure that is much quicker at reaching decisions and enables people closest to the work to drive improvements. This does not imply that all of the management team who achieved their positions through subject matter expertise are obsolete. It simply implies that the skills required to manage the cross-functional efforts of this evolving organization are not those that historically dictated management selection. Leadership will be more about the ability to encourage the workforce and focus the team on waste reduction, and then intimacy with the historic performance or knowledge of previous issues. There is the potential for two independent streams of management capabilities to evolve:

- **Dedicated continuous improvement teams**—Some organizations will gravitate to implementing continuous improvement teams and the management of them as a core functional capability. The thought processes is that while every employee will be responsible for knowledge of the tools and processes used in Lean, some will naturally be better at seeing the waste and eliminating it than others. By congregating these highly capable individuals into a task force to attack major issues and provide mentoring and coaching for the masses, you end up with a hierarchy of continuous improvement capabilities. The risk here is to ensure that enough evidence of demonstrative capabilities in learning exists at the individual contributor level to show that the organization is in fact evolving at an acceptable rate of learning. In other words, you have to be careful that the segregated group driving improvements is not responsible for all of the results; this cannot be an elitist solution.
- **Cross-functional self-directed teams**—As it relates to the individual contributors, it is important to recognize that the strategy for cross-functional solutions has become the norm. These teams will demonstrate robust process controls, systemic approaches to continuous improvement, and exemplary management of process escapes. It is how the team treats the process escapes that you will get further insights into their own overall capabilities. By now, we want the organization to recognize error escapes deviations in all forms are gems that need to be guarded and treated as precious. These gems provide us with meaningful examples of how the processes we are using fail and as such offer

the clearest insight into the opportunities for improvement. By watching how the team addresses and responds to these items you will determine whether the passion for finding, protecting, and leveraging these opportunities is at the level you feel it should be for a continuously learning organization. Another indicator of success will be the balance between external and internal verification capabilities and the performance metrics and monitoring of the processes. Here the traditional manager who is an SME on one dimension of the process will act as a coach and mentor to the evolved team, sometimes taking on the role as the evaluator of their performance, sometimes as a contributor to the solution.

7.2 Methods

In Chapter 5, we described the shift in focus for the organization from performance metrics to process control and process development. A bellwether of your success will be the degree to which the organization is establishing industry-leading processes and the tools required to deliver the solutions. Built in-house should become the norm for many of the solutions that you require. This is an indicator that you are outpacing the service provider's design capabilities with the solutions that you are deploying.

As you look at the landscape of how these methods have influenced the evolution of the bulkier processes, you should now start to see how these migrate into other areas of the business that are historically not prone to this type of improvement. New product interaction and mergers and acquisitions activity are examples of areas where the disciplines in the process improvements associated with Lean have tremendous potential, and where creating a Lean Learning Organization can truly lead to a sustainable competitive advantage.

We would encourage you to be thinking now at the most holistic level of the processes used to determine where to focus the organization, and how to rank and stack the opportunities for improvement. Recognize that perfection is not the goal and subsequently the area that you focus your improvements on should be characterized in a way that allows you to assess internally or externally when it is good enough.

As we have seen throughout the book, LWL is not a formula, it is not a canned approach, nor is it the only game in town. We have shown that one of the key attributes of LWL is its expectation that individuals, teams, and the whole organization will openly share the methods and systems used to achieve both successes and failures.

Sharing successes is a natural reaction for most people. They want others to know what challenge they were faced with, how they determined how to solve it, how they put the solution in place, and how the problem was fixed. There is a strong desire because of recognition for a job well done and challenges overcome. It takes almost no effort to solicit the sharing from successes. It is a great time when organizations, teams, and individuals can do that. We encourage you to celebrate these successes.

The opposite is true at all levels for learning through failures. By nature, failures represent a negative impact on a challenge or opportunity. It said something went wrong, that a solution was ineffective, or a problem was poorly understood and the solution was the wrong answer. It is not natural for people to talk about their failures or what they learn from them. This is where the energy must be spent by leaders and management to develop approaches that foster, and even promote, the lessons from failures. Colin Powell, U.S. Army General and Statesman, said that learning from failures is the greatest level of learning.

> There are no secrets to success. It is the result of preparation, hard work, and learning from failures.
>
> **Colin Powell**

In truly successful organizations that embrace the art of learning, there are methods or approaches that make learning from failures relatively painless. As senior leaders, you must set the culture that says learning through failure is okay. The idea of noisy learning, noisy failures, is one that is rooted in the art of decomposing failures and searching for nuggets of learning.

7.3 Machine

There should be numerous examples visibly evident to the organization where the tools used to manage the business have changed. Common deployment of MSW should make the process capabilities and improvement focus of each critical stream in the process visible to those doing the work. Management language used in discussion of these energies and efforts will be consistent across the organization with the team relying on expertise as required based on the projects or problems they are trying to solve.

In other words, the tools of problem solving will be universally deployed. Just as not every employee is omnipotent in all dimensions of his or her work, experts will be identified in certain dimensions of problem solving capabilities. These experts will be leveraged across the organization if, and when, they are needed. The vocabulary of the organization will have changed, and it is in this common language that you will see some of the aspects of OL evident.

As you look back on how you established the performance of the group to this stage, it should be very clear that the path that you followed was the result of thousands of individual decisions and developments made during the journey. We will go one step further to say that given the exact same set of circumstances, it would be virtually impossible for you to recreate verbatim every improvement project, every learning opportunity, every engagement that guided the organization to the point it is today. This is, in fact, the strength of the strategy. It is obvious at this stage that the capabilities the organization has worked hard for, fought for, and won are truly unique to those decisions and learning opportunities that were created through the journey. Think of trying to have a consultant come in and implement this in a short period of time; the likelihood of success is very low. You cannot buy what you have just established. You had to build it, and in doing so, you created a structure that is not only unique, but also is very strong.

Further indicators of success will simply come from things like people wanting to work in your organization. Because the level of investment that you are making in the evolution of people's capabilities, it will become obvious to everyone that they are better or more valuable employees today than they were yesterday. This will have a great appeal not only to the staff that you try to hire but also to those that you retain. You, your staff, and your organization's performance will be the best recruiting tool that you have.

7.4 Final Thoughts from the Authors

As you think about the next stage of your journey, we recognize that nothing is static. No matter how good you are performing, you will be faced with new challenges be they new products, customers, markets, or threats from disruptive technologies. The difference having created a culture of LWL is that you now have an organization that understands and believes in the capabilities it has. You can face these challenges with the confidence that comes from competence. This is not organizational arrogance, but rather a team that understands that they have the capabilities to be competitive on any stage, on any dimension, and it is with that confidence and enthusiasm that you can take on new challenges with gusto.

We hope that you enjoyed the journey that you have undertaken with your group. Please reach out and share with us some of the successes, failures, and learnings that you discovered along the way.

Key Messages in This Chapter

1. The organization has attained critical mass and is improving without incremental investment from you. Be careful that you don't get left behind! Make sure that you personally continue to drive improvements in the work that you and your team do….i.e. your MSW.
2. The organization has flattened and is much more nimble than ever before. How are you leveraging this capability? Are leaders rotating through function roles? Different sites?
3. Failures offer the greatest opportunities for learning. How are you capturing these gems so the team can learn from them?
4. Your team will be poached…plan for it?
5. The system you have deployed is repeatable…how are you using that to shape your organic growth and acquisition strategy?

Challenge Actions

This section is about wrapping up what you've accomplished and learned in this wave of transformation. It is a reflective period where you personally need to summarize how you have transformed, what was done right and what you would do differently the next time.

Choose actions totaling at least 50 points, ones that help you and your organization understand the transformation that has occurred and how it could be tuned.

Challenge Actions—Leverage the Learnings	*Points*
1. Are you as good as you could be? Take the time to ensure thee the rate of development in the last holdout areas is now meeting your expectations. There should be sufficient Credibility Cash form the entire organization to move just about any obstacle. If there are a few senior leaders still holding on the way it used to be...make the change!	30
2. If you are performing at this level you are now the benchmark reference. Are academics and industry leaders spending time in your business? How do you balance these requests and demands on your team's time with the daily workload?	25
3. Has your career path planning and other HR functions kept up with the new dynamics of your workforce? What has changed in your recruitment, hiring, orientation and training programs? How has your compensation structure adjusted to recognize Lean talents and learning capabilities?	25
4. What would your LWL plan look like for the next acquisition? How would you accelerate the transformation?	20

Glossary of Terms

3 Ms (Man, Methods, Machines): Shorthand approach to using a fishbone analysis method with just three of the typical buckets used to capture classes of contributing factors.

3 Modern Wastes: *See* **Muda**.

5 Ms: Typical bones of a Fishbone diagram.

Man, Methods, Materials, Machines, and Management: *See* **Fishbone diagram** or **Ishikawa diagram**.

5 S: Five aspects of using Lean to improve the workplace: Sort, Set in Order, Shine, Standardize, Sustain.

5 Whys: Approach of asking "Why?" five times in a row to help find the root cause.

6 Sigma: Six Sigma is an organizational management system that seeks to improve the quality of process outputs by identifying and removing the causes of defects (errors) and minimizing variability in manufacturing and organization processes.

7 Traditional Wastes: *See* **Muda**.

8D report: Systematic 8-step approach to root-cause analysis and permanent correction.

A3 report: Improvement project report format; everything is on a single page.

Applied Rate of Learning (AROL): Actions or activities that engage an organization and its employees directly with a phenomenon (condition, opportunity, or challenge) being examined and involve them in creating solutions that connect the phenomenon with application of theoretical concepts.

As-is state: Accurately captured condition, attributes, and execution of a process as it is running today. *See* **Future state** or **Should-be state**.

Benchmark: A specific output, deliverable, or attribute measure that is used for comparison.

Benchmarking: The act of comparing outputs, deliverables, or attributes of a product, service, or process that yields one or more benchmarks.

BKM: Best known methods.

Capability index: A degree of process margin around a defined, in-control process (comparison of mean ± 3 sigma versus specifications/requirements) (*see* C_p, C_{pk}).

CCI: *See* **Credibility Cash Index**.

Charter: Formal document that creates an improvement team and sets the project work foundation.

Change-hungry organization (CHO): Organization's traits that lead to ever-present drive to seek, find, and solve opportunities to improve performance as measured in time, cost, quality, and satisfaction.

CIP: Continuous improvement process.

CMMI: Capability maturity model integrated.

Control chart: Statistically derived chart with data, control limits, specification limits, and descriptive statistics.

Control limit: Statistically calculated process limits based on variation in a data set.

Correlation analysis chart: A mathematical technique that is used to study changes in measured signals.

Credibility Cash Index (CCI): An index combining measures of knowledge, capability, historic success, energy to learn and change, and ability to share and teach.

DFSS: Design for Six Sigma. *See* **DMADV**.

Dissonant technology: The state of holding two or more conflicting cognitions (e.g., competing technologies, markets, strategies) simultaneously in a state of dissonance.

DMADV: Define, Measure, Analyze, Design, Validate—5 steps in Design for Six Sigma.

DMAIC: Define, Measure, Analyze, Improve, Control—5 steps in using Six Sigma. *See also* **6 Sigma**.

DoE: Design of Experiments.

EECO: Effective efficient consistent organization—the three measurable attributes of organizational performance that can be applied to the execution or output of any process or system.

Entropy: A measure of how evenly energy is distributed in a system; in a physical system, entropy provides a measure of the amount of energy that *cannot* be used to do work.

Error proofing: Methodology to build failure prevention into the design of products, services, and processes. *See* **Poke yoke**.

Fishbone diagram: Graphical technique to brainstorm or analyze contributing factors for a problem or opportunity. *See* **5 Ms** or **Ishakawa diagram**.

Fitness-for-use: Effectiveness of a design, manufacturing method, and support process employed in delivering a good, system, or service that fits a customer's defined purpose, under anticipated or specified operational conditions.

Future state: Proposed condition, attributes, and execution of a process as it will be running at a prescribed time in the future (post improvement). *See* **As-is state** or **Should-be state**.

Gemba walk: Process of walking the actual process flow from beginning to end while recording steps, rework loops, wait times, work times, inventory between steps, and more. *See* **Walk-the-flow**.

Goodness-of-fit: A statistical model describing how well a distribution fits a set of observations.

Histogram chart: Graphical representation of data, where similar data are grouped into cells.

In-control processes: Defined by the analysis of process attributes when compared to a set of control rules. While many rule sets exist, the Western Electric rules are typically used. *See* **Out-of-control processes**.

Ishikawa diagram: Graphical technique to brainstorm or analyze contributing factors for a problem or opportunity. *See* **5 Ms** or **Fishbone diagram**.

Johari window chart: Graphical representation of data that combines three measures. It uses an X-Y graph plus a third value that affects the size of each data point in the graph.

Judoka: Process abnormalities (escapes).

Kaizen: An improvement action, usually of short duration time, that identifies a problem, defines root cause, and implements a solution for permanent correction.

KISS: Keep It Simple Stupid (variations include "keep it short and simple," "keep it simple sir," "keep it simple or be stupid," "keep it simple and stupid," "keep it simple and straightforward," and "keep it simple and sincere").

Lean: A production practice that considers the expenditure of resources for any goal other than the creation of value for the end customer to be wasteful, and thus a target for elimination. Working from the perspective of the customer who consumes a product or service, "value" is defined as any action or process that a customer would be willing to pay for.

Lean thinking: Changes the focus from optimizing separate technologies, assets, and vertical departments to optimizing the flow of products and services through entire value streams that flow horizontally across technologies, assets, and departments to customers.

Learning atrophy: An organization's or individual's loss of ability to probe and ask why, discover new methods, and be able to communicate what has been discovered so others can apply and improve quicker.

Learning with Lean: Using Lean tools, methods, and disciplines to create an environment where asking why, seeking data and information, and then combining findings to remove wastes. *See* **Minimalistic process design**.

Learning organization (LO): An organization that enables learning of its members and continuously transforms itself; it has five main features: systems thinking, personal mastery, mental models, shared vision, and team learning. *See* **Organizational learning**.

Mean: Mathematically the same as a data set average.

Minimalism: Describes movements in various forms of art and design, especially visual art and music, where the work is set out to expose the essence, essentials, or identity of a subject through eliminating all non-essential forms, features, or concepts.

Minimalistic (or minimalist) process design: Borrowed architectural practices that reduce process construction to the barest, most functional, most efficient assembly of the fewest steps with near zero embedded waste. *See* **Muda** and **5 S**.

MSA: Measurement system analysis.

MSW: Management standard work.

Muda: A traditional Japanese term for an activity that is wasteful and does not add value or is unproductive.

Necessary and sufficient test: A simple test that examines actions versus objectives and drives a conclusion on the right type and number of actions required for success of the objective.

New product introduction (NPI): Designing, fabricating, and releasing a new product or service for sale or use through a structured process (typically program management).

Non-value add (NVA): A process step identified in a VSM that does not add value to the deliverables and does not pass the typical customer litmus test—the customer is not willing to pay for execution of the step. *See also* **VA, VE, VSM**.

OL: *See* **Organizational learning**.

One-touch flow (OTF): A process design where only the fewest value-add steps exist and each step is done with a single touch.

Operational performance quotient (OPQ): The quotient is an integration of multiple measures (scores) of operational performance using a weight score arriving at a single value suitable to determine the health or performance of the organization.

Organizational learning (OL): An area of knowledge within organizational theory that studies models and theories about the way an organization learns and adapts; one of its focus areas is the process of detection and correction of errors. *See* **Learning organization**.

OTF: *See* **One-touch flow**.

Out-of-control process: Resulting determination using various rules for statistical trends in run charts. *See* **In-control process**.

Pareto chart: Graphical analysis of data that is classified in sub-groups and sorted for highest occurrence (generally left-most referenced in the chart).

PDCA: Plan-Do-Check-Act, also called the Deming Cycle or Shewart Cycle.

Performance measures: Classes of measures used to gauge performance: Time, Cost, Quality, Satisfaction, Continuous Improvement (Learning), and Governance.

Pissed-off value of marginal performance (PVMP): Representation of a customer's experience where many to most of the delivery, relationship, and follow-through actions are performed at marginal to mediocre levels where the customer would complain for any single action, but in total the overall customer experience is negative.

Poke yoke: Methodology to build failure prevention into the design of products, services, and processes. *See* **Error proofing**.

Predictive measures: Set of process measures that monitor an attribute of the process. The resultant measure can be used to predict a condition sometime in the future. *See* **Reflective measures**.

Probability: An attitude of mind toward some proposition of whose truth is not certain.

Process dust: *See* **Techno-waste**.

PVMP: *See* **Pissed-off value of marginal performance**.

Queue time: Measurable time between process steps or actions where a deliverable is sitting or waiting for the next action to occur. Generally, no value-add action is happening to the deliverable during this time.

Rack, stack, and select: Simple task of collecting gaps or opportunities (racking) and then placing them in prioritized order (stacking) using a set of prioritization criteria.

Rate of learning: Represented as the change in demonstrated skills (under analysis) over time; typically, it takes on an "S"-shaped curve.

RCA: Root cause analysis.

RCCA: Root cause corrective action.

Reflective measures: Set of process measures that monitor the historic performance of an attribute of the process. *See* **Predictive measures**.

Return on Investment (ROI): Financial measure of investment performance.

RPN: Risk priority number. *See* **FMEA**.

Should-be state: Proposed condition, attributes, and execution of a process as it will be running at a prescribed time in the future (post-improvement). *See* **As-is state** or **Future state**.

SIPOC: Supplier–Input–Process–Output–Customer.

SME: Subject matter expert.

Standard deviation: The probability of obtaining a test statistic at least as extreme as the one that was actually observed, assuming that the null hypothesis is true.

Swim lane: A segmented partition on a VSM generally assigned to either a type of work or a function that owns work being done. *See* **VSM**.

SWOT analysis: Analytical process to find Strengths, Weaknesses, Opportunities, and Threats, and determine their impact on the mission success of the organization.

Takt: Measure of a process's beat rate when output expectation is known and available time and resources are known.

Techno-waste: Non-value steps in an automated process that offer almost no savings in time if removed and, by leaving them in the process, foster a behavior opposite of OTF. Also called *process dust*. *See* **One-touch flow**.

Try-storming: Integration of brainstorming and simulation through rapid construction of a prototype solution (out of anything that is available) allowing people to visualize, touch, and further improve on an initial idea.

Value Add (VA): A process step identified in a VSM that adds value to the deliverables (typical customer litmus test: the customer is willing to pay for execution of the step). *See* **NVA, VE, VSM**.

Value Enable (VE): A process step identified in a VSM that does not directly add value to the deliverables, but is required (e.g., by regulation, law, organizational policy, or others); it does not pass the customer litmus test. *See* **VA, NVA, VSM**.

Value stream map (VSM): Graphical depiction of a prescribed process using a flow-like pattern partitioned into swim lanes, supported with data showing work duration and idle time (queue time), rework loops, measures of success, and others. *See* **Swim lanes, NVA, VA, VE**.

Variation: A normalized measure of dispersion of a probability distribution.

Voice of the Customer (VoC): The system of logically segmenting customers into classes, intelligently asking for and receiving feedback, distilling and extracting key information and messages from the feedback, and focusing the information and feedback into significant guidance for changes to deliverables (product, service, and relationship) received by the customers.

VSM: *See* **Value stream map**.

Walk-the-flow: Process of walking the actual process flow from beginning to end while recording steps, rework loops, wait times, work times, inventory between steps, and more. *See* **Gemba walk**.

Waste walk: *See* **Walk-the-flow** and **Gemba walk**.

Appendix

Tools and Templates

Contained in this section are various assessment, decision-making, and analysis tools discussed throughout this book. Their descriptions, construction, scoring, and score interpretations precede exhibits of the full instrument. In Table A.1, each of the instruments is cataloged and provided with a URL reference where you can acquire a copy for your direct use or adaptation to tailor it specifically to your organization.

Table A.1 Assessment Instruments

	Instrument	*URL Reference (read-only)*
Learning Organization *(Section A.1)*	Leading a Learning Organization assessment	
Lean Leadership *(Section A.2)*	Lean Leadership assessment	
Lean Thinking and Acting *(Section A.3)*	Lean Thinking and Acting assessment	
Credibility Cash Index *(Section A.4)*	Credibility Cash Index 360-degree Inventory	

A.1 Assessment—Leading a Learning Organization

Description

This self-assessment instrument uses 10 questions focused on the behaviors and actions of organizational leaders responsible for guiding learning organizations.

Purpose

Its purpose is to cover a reasonable range of senior level behaviors that demonstrate their understanding of what a Lean organization looks and behaves like.

Where Used

This instrument can be used across any type of organization in any geographic or regional site, but is targeted at the leader and executive management levels.

When Used

Typically, this instrument is used at the onset of an assessment period when the as-is condition is unknown, or at some periodic rate when the leaders choose to refresh their knowledge of how well they are demonstrating good practices that advance become a Learning organization.

Construction

The instrument uses 10 questions segmented into three roles: designer, teacher, and steward. There are three questions each for designer and teacher, and four questions examining a leader's stewardship behaviors.

The tool is constructed so that the scores for each individual question are plotted by role so that the scores can be interpreted at the individual level. The tool can be adapted such that scores from multiple management members can be integrated to form a collective picture using the same graphs. If this is done, it provides the ability to analyze individuals against group norms.

Scoring Range and Definitions

Each question can be scored using these definitions:

- **Little or no evidence of leader behaviors and actions (score = 1)**—There is no evidence that leadership or management is adopting or promoting organizational learning practices, methods, or tools. There is demonstrated aversion or rebuff to practices by teams or individual contributors.

- **Some ad hoc evidence of leader behaviors and actions (score = 2)**—Management passively accepts, but does not endorse, the use of organizational learning methods and tools by teams and individuals. The leadership members do not personally use methods to manage the organization.
- **Occasional or often evidence of leader behaviors and actions (score = 3)**—Management passively accepts and endorses the use of organizational learning tools and methods across the organization. They set limited expectations for its use as part of normal activities. There is some evidence of management's use of the tools and methods to standardized work activities and outputs. No systematic learning is in place at this level.
- **Most of the time there is evidence of leader behaviors and actions (score = 4)**—Proactive use of learning tools and methods across the organization and by management is evident. Some evidence of becoming a Learning organization is becoming a normal practice in work areas, and learning systems to capture and proliferate key lessons are in the early stages of introduction and use.
- **Consistent or always evidence of behaviors from the leader (score = 5)**—Management has adopted and is practicing being a Learning Organization. They are seen by the organization as key drivers to cultural and behavioral changes. Learning systems are in place and proactively driving cycles of learning and proliferation of key lessons. Management methods are now systematic in nature with clear expectations of execution. Integration of learning objectives is part of MSW and is now a norm for all levels of management.

Scoring Interpretation

Scores from this instrument are initially divided between the three roles: designer, teacher, and steward. While not part of the construction of the assessment tool, there are considerations based on the maturity of the organization and where it is and its Lean implementation journey. Each of the roles may be more important than others. The tool as it is constructed does not represent that, nor does it provide additional waiting for those important areas. Organizations using this assessment tool should evaluate if this condition exists and tailor both the instrument and its scores accordingly.

Exhibits

This tool can be accessed at: j.zurn@hotmail.com

Assessment — Leading a Learning Organization
Scoring:
Little/no evidence of leader behaviors/actions
Some/ad-hoc evidence of leader behaviors/actions
Occasional/often evidence of leader behaviors/actions
Most of the time there is evidence of leader behaviors/actions
Consistent/always leader behaviors/actions

	Leader Attributes — Learning Organization	Scores					**Score by Role**
		1	2	3	4	5	
Designer	Provide the 'end game', direction and speed for change				×		4.0
	Sets clear expectations supported with policies and strategies that align and enable individual and organizational learning to occur					×	
	Is a content expert, unique to the organization, with the ability to match learning styles and processes with organizational culture and behaviors, while targeting appropriate impacts to the organization			×			
Teacher	Presents themselves as a knowledge leader and demonstrated practitioner of Lean skills			×			3.7
	Is able to effectively translate Lean technologies across all functions, and up or down the leadership chain			×			
	Can effectively convince and enroll lay people employees, who have new or emerging skills, for the purpose of continuous improvement systems					×	
Steward	Seen as the "lightening rod" for the Lean transformation			×			3.3
	Has the passion and the drive to fully engage with improvement teams, regardless of level				×		
	Is capable of defending the approach in deployment against natural naysayers			×			
	Maintains strong presence, visibility, and communication actions that ensure successful deployment and long term sustainability			×			

Exhibit A.1 Leading a Learning Organization.

A.2 Assessment—Lean Leadership

Description

This self-assessment instrument uses self-scoring of 38 questions that focus on Lean leadership approaches in project implementation and daily operations.

Purpose

Its purpose is to quickly evaluate a wide range of expected behaviors using evidence, artifacts, or system improvements.

Where Used

This self-assessment instrument can be used across any type of organization, and in any geographic or regional site. The questions are structured, and it is intended to be used by any level of team leader, supervisor, manager, or executive.

When Used

Typically, this instrument is used at the onset of an assessment period when the as-is condition is unknown, or at some periodic rate when the organization chooses to refresh its understanding of current levels of Lean behaviors after change activities have been put in place.

Construction

The instrument uses 38 questions segmented into two areas of accountability: project implementation and ongoing operations. There are 16 questions covering project implementation, and 22 questions examining ongoing operations.

The questions are assembled using eight attributes for each area of accountability. These include:

- passion for Lean
- disciplined adherence to process accountability
- project management orientation

- Lean thinking
- ownership
- tension between applied and technical
- balanced commitment to production and management systems
- effective relations with support groups

The tool is constructed so that the scores for each individual question are plotted by area of responsibility so that the scores can be interpreted at the individual level. The tool can be adapted such that scores from multiple management members can be integrated to form a collective picture using the same graphs. If this is done, it provides the ability to analyze individuals against group norms.

Scoring Range and Definitions

Each question can be scored using these definitions:

- **N/A (score = 0)**—Not applicable.
- **Little or no evidence, avoidance of Lean (score = 1)**—There is no evidence that leadership or management is adopting Lean practices, methods, or tools. There is demonstrated aversion or rebuff to the use of Lean practices by teams or individual contributors.
- **Some ad hoc evidence, slow adoption (score = 2)**—Management passively accepts the use of Lean tools and methods by teams and individuals, but it does not proactively drive its adoption by the organization nor personally uses Lean leadership methods to manage the organization.
- **Some systematic evidence, no learning (score = 3)**—Management passively accepts and openly endorses the use of Lean tools and methods across the organization and has set expectations for its use as part of normal activities. There is some evidence of management's use of the tools and methods to standardized work activities and outputs. No systematic learning is in place at this level.
- **Lean systems evident, few cycles of learning (score = 4)**—Proactive use of Lean across the organization and by management is evident. Standard work is becoming a normal practice in work areas, and learning systems to capture and proliferate key lessons are in the early stages of introduction and use.

- **Lean systems and behaviors are key drivers (score = 5)**—Management has adopted and is practicing Lean management techniques and is seen by the organization as a key driver to cultural and behavioral changes. Learning systems are in place and proactively driving cycles of learning and proliferation of key lessons. Management methods are now systematic in nature with clear expectations of execution. Management standard work is now a norm for all levels of management.

Scoring Interpretation

Scores from this instrument are initially divided between project implementation and ongoing operations. Each of these is graphically seen in the polar diagrams on the first page of the instrument.

Common to project implementation and ongoing operations scores, the desired representation of the scores should be a balanced circle of scores ranging between 3 and 5. Some variation is expected at the initial time of use of this instrument due to early maturity levels, familiarity with the tool, or calibration of scoring criteria.

Again and during both project implementation and ongoing operations the desired state, when using the tool over a period of cycles, is to see growth in the circle for each assessment result.

Unique to ongoing operations, there are 50% more questions and, at the initial stages, even greater variation is expected than the score set for project implementation.

While not part of the construction of the assessment tool, there are considerations based on the maturity of the organization and where it is and its Lean implementation journey. Some of the eight areas may be more important than others are. The tool as it is constructed does not represent that, nor does it provide additional waiting for those important areas. Organizations using this assessment tool should evaluate if this condition exists and tailor both the instrument and its scores accordingly.

Exhibits

This tool can be accessed at: j.zurn@hotmail.com

Check – Lean Leadership Self Assessment

Purpose: Assess Leadership's Lean Culture and Competency
Scoring: 0 N/A – cannot score
1 Little/no evidence, avoidance of Lean
2 Some/ad-hoc evidence, slow adoption
3 Some systematic evidence, no learning
4 Lean systems evident, few cycles of learning
5 Lean systems and behaviors are key drivers

Color Code for Spider Charts (below)

- Passion for lean
- Disciplined adherence to process accountability
- Project management orientation
- Lean thinking
- Ownership
- Tension between applied and technical
- Balanced commitment to production and management systems
- Effective relations with support group

Lean Leadership for Project Implementation

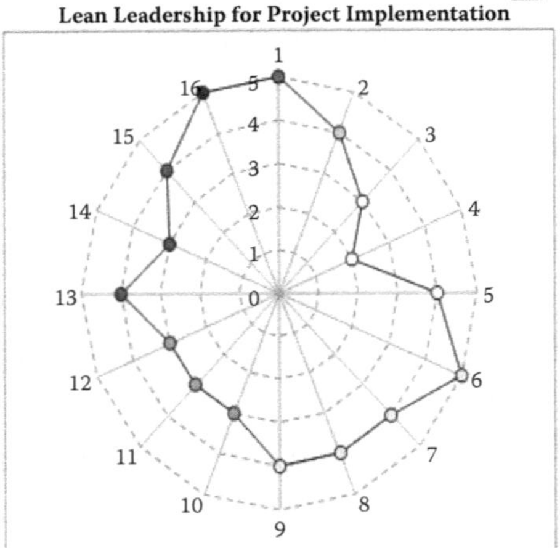

Lean Leadership for Ongoing Operations

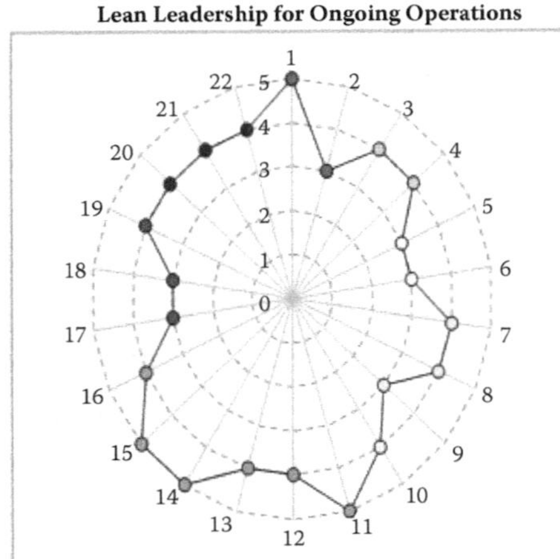

Attribute	For Project Implementation	0	1	2	3	4	5	For Ongoing Operations	0	1	2	3	4	5
Passion for Lean	1. Passionate about the potential for Lean to make the enterprise more successful and work more fulfilling for all involved.						×	1. Passionate about the potential for Lean to make the enterprise more successful and work more fulfilling for all involved.						×
								2. Willing to make personal changes in one's own work, including using standardized work for his/her own position.				×		
Disciplined adherence to process accountability	1. Sets expectations, regular uses a process to track and follow up on actual accomplishment of assigned tasks.					×		1. Sets expectations, regular uses a process to track and follow up on actual accomplishment of assigned tasks.					×	
								2. Exhibits intense commitment to focus on explicitly defining processes and disciplined adherence to them.					×	
Project management orientation	1. Prior experience in successfully implemented projects.				×			1. Able to identify needed changes based on daily process data and assign small-bite daily tasks leading to successful implementation of the changes.				×		
	2. Uses a defined process to track performance and completion of task assignments.			×				2. Uses explicitly defined visual processes to track and follow-up assignments and take appropriate corrective action.				×		
	3. Identifies corrective action where necessary and follows up on it.					×								

Exhibit A.2a Lean leadership.

Lean thinking	1. Understands Lean concepts.						×	1. Serious about ongoing improvement based on a goal of perfection.					×	
	2. Has had experience applying Lean concepts.					×		2. Sees with "kaizen eyes."					×	
	3. Talks about and promotes a Lean future state.					×		3. Holds and coaches a root-cause orientation to corrective action.				×		
	4. Finds ways to apply and illustrate Lean concepts in daily project work processes.					×		4. Has learned process improvement/problem solving methods; able to personally lead Lean process improvement.					×	
Ownership	1. Thinks and talks about the area as his/hers to lead, set direction for, change, and improve.				×			1. Thinks and talks about the area as his/hers to lead, set direction for, change, and improve.						×
								2. Eager to empower others in the area through structured ways to elicit and implement their ideas.					×	
								3. Acknowledges and celebrates improvements made by others at all levels.					×	
Tension between applied and technical	1. Understands the need to sweat the details, as well as to get things done.				×			1. Understands and respects the details behind elements of Lean, such as flow, pull, standardized work, etc.						×
	2. Willing to listen to technical experts and consider their advice in planning for the implementation.				×			2. Actively supports steps to upgrade performance and expose previously hidden impediments.						×
								3. Takes a "what can we do today" orientation to making change happen steadily, step-by-step.					×	
Balanced commitment to production and management systems	1. History of effective give and take with people and ideas.					×		1. Personally treats process focus as crucial to the area's success; is able to see waste and opportunity even in Leaner processes.				×		
	2. Evidence of process focus beyond a "hit the numbers" approach to management.				×			2. Insists on compliance with requirements for visually tracking process performance and execution.				×		
	3. Eager for greater participation by production people as well as others.					×		3. Insists on analysis and appropriate, timely action on impediments to normal operation of process.					×	
Effective relations with support groups	1. History of getting things done with support from operations support groups such as engineering, quality, production control, safety, finance, HR.						×	1. Understands roles, responsibilities, and expertise of support groups.					×	
								2. Incorporates support groups appropriately in plans for improvement and responses to problems.					×	
								3. Makes expectations explicit for support group performance in support of production processes.					×	

Exhibit A.2b Lean leadership.

A.3 Assessment—Organizational Lean Thinking and Action

Description

This self-assessment instrument uses self-scoring of 24 questions that focus on personal approaches to Lean thinking, decision making, and action taken.

Purpose

Its purpose is to quickly evaluate a wide range of situational conditions and help determine how closely your responses align with the desired behaviors.

Where Used

This self-assessment instrument can be used across any type of organization, at any employee level, and in any geographic or regional site.

When Used

Typically, this instrument is used at the onset of an assessment period when the as-is condition is unknown, or at some periodic rate when the organization chooses to refresh its understanding of current levels of Lean behaviors after change activities have been put in place.

Construction

The instrument uses 24 questions, each of which are scored independently, ranging from strongly disagree to strongly agree (score range is from 1 to 5). These questions are topically collected into six pattern areas that include:

- resource maximization
- customer value
- systematic problem solving
- effective decision making
- knowledge capture and reuse
- support to grow problem-solving skills

The tool is constructed so that scores can be interpreted at the individual level or at a collective summary level when many employees combine their scores. Based on scores of the six pattern areas, there is a series of recommended actions to assist in targeting improvement areas.

Scoring Range and Definitions

Each question can be scored using the following definitions (note: there is no opting out of the scoring by using N/A or 0 on this assessment; a position and score is required for all questions):

- **Strongly disagree (score = 1)**—Definitively disagrees with the premise of the question and all of its components. This represents an adamant negative position.
- **Disagree (score = 2)**—Partially disagrees with the question's premise, or disagrees with some of the components of the question. The respondent is more negative than neutral, but it is not an adamant negative position.
- **Neutral (score = 3)**—Represents a range of responses that can include equal negative and positive agreements to the premise and its components, middle of the road agreement that the premise and its components are being done adequately according to needs, or an ambivalent position on the question due to a wide range of potential causes.
- **Agree (score = 4)**—Partially agrees with the question's premise, or agrees with some of the components of the question. The respondent is more positive than neutral, but it is not an adamant positive position.
- **Strongly agree (score = 5)**—Definitively agrees with the premise of the question and all of its components. This represents an adamant positive position.

Scoring Interpretation

For the purpose of interpretation, the 24 questions are collected into six pattern areas:

- Resource maximization: questions 6, 8, 24
- Customer value: questions 5, 12, 21

- Systematic problem solving: questions 4, 7, 11, 16, 19
- Effective decision making: questions 1, 10, 13, 15, 23
- Knowledge capture and reuse: questions 3, 9, 18, 20
- Support to grow problem-solving skills: questions 2, 14, 17, 22

Exhibits

This tool can be accessed at: j.zurn@hotmail.com

Check – Lean Thinking & Action Self-Assessment

Note: Place an X in only ONE score box for each question (light yellow cells below).

	Question	Strongly Disagree 1	Scores (*Enter score in ONLY one*) Disagree 2	Neutral 3	Agree 4	Strongly Agree 5
1	When we go into a decision meeting or project review, we know that the right people will be in the room to make decisions.					
2	Managers in my organization ask good, challenging questions when I present my findings and recommendations to them.					
3	When I recognize a problem that we have seen before, I have the ability to find out how we solved it last time.					
4	We explore multiple alternatives before making key decisions.					
5	Subject matter experts (SMEs) get regular opportunities to deepen their customer knowledge with early customer feedback on their process designs.					
6	Process execution and reviews meet their schedules consistently.					
7	Decision makers take the time to understand the problems, alternatives and recommendations before making a decision.					
8	There are few loopbacks late in an improvement action's implementation cycle.					
9	We take the time to capture what we've learned so that we can share it with others and reuse it ourselves later.					
10	We do not hold up the flow of improvement work with artifical barriers imposed by reviews or decision meeting.					
11	We take some time to understand root causes before we recommend countermeasures or solutions.					
12	Customers are highly satisfied with our improvement action's quality.					
13	When we go into a decision meeting or review, we are confident that there will be no last minute objections or issues to delay decisions.					
14	Managers in my organization use systematic problem-solving to solve problems within their span of control.					
15	We take the time to identify the people who need to be consulted about a decision, and then engage them in the decision-making process.					
16	We take the time to measure results and reflect upon the effectiveness of the decisions that we make so that we can learn.					

Exhibit A.3a Lean thinking and action.

17	Our group's leaders do not second guess the decisions they have delegated to their teams.					
18	My team's information systems make it easy to find documentation from past projects that could be helpful to me now.					
19	We understand what problems we need to solve early in the improvement cycle to minimize implementation risk and we solve them before we release an improved process.					
20	We actively search out reusable knowledge and expert input as part of our problem-sloving and decision-making processes.					
21	We know what value-add work our customers will 'pay for' and we have the ability to modify our processes to achieve time, cost and quality targets.					
22	We know who has the authority to make the important decisions in our organization.					
23	We have the ability to kill an improvement project in early development when it becomes clear that the project won't meet its goals.					
24	We have the capacity to deliver process improvement needed to achieve our organization's growth targets without overloading resources or adding additional staff.					

What Patterns Did you Observe?

Score 1 point for each Strongly Disagree, 2 for Agree, 3 for Neutral, 4 for Agree and 5 for Strongly Agree (Automatically calculated — DO NOT ENTER A SCORE)	Max Pts.	Score:	
Questions 6, 8, 24: Resource Maximization: Ability to optimize capacity, improvement resource utilization and achieve improvement schedules.	15	6.8	Medium
Questions: 5, 12, 21: Customer Value: Ability to maximize customer value, including benefits, quality and cost.	15	8.8	Medium
Questions 4, 7, 11, 16, 19: Systematic Problem Solving: Ability to use problem-solving methods that solve problems systematically and permanently.	25	14.1	Medium
Questions 1, 10, 13, 15, 23: Effective Decision-Making: Ability to make decisions that deliver the expected results without being revisited or causing unintended side effects.	25	15.8	Medium
Questions 3, 9, 18, 20: Knowledge Capture and Re-use: Ability to maximize value from the organization's knowledge.	20	12.7	Medium
Questions 2, 14, 17, 22: Support to Grow Problem Solving Skills: Ability to support each other and coach staff to build problem-solving ability in every member of the team.	20	13.0	Medium

Exhibit A.3b Lean thinking and action.

A.4 Inventory—Credibility Cash Index

Description

The CCI inventory tool is used to gather peer, subordinate, and manager input on five critical elements senior leaders need to be able to effectively kick off, personally invest, and lead organizational change.

Purpose

The tool is designed to subjectively gauge the amount of personal credibility that is at your disposal to engage with the organization and cause a meaningful change. Using a gap analysis between the level of CCI inventory the leader possesses and the amount of required to produce the AROL provides insight into the difficulty for change, level of organizational stress required for movement, and where the CCI will be applied. The total energy required to move the organization can come from external stress, your CCI, the application of others CCI, and or the focused combination of all of these.

Where Used

It is oriented for the senior leader and their staff leaders. It has the potential to be tailored and used to evaluate a Lean leader's ability to influence his or her local work teams.

When Used

The CCI inventory tool should be used at the onset of a significant change initiative and it should become a routine tool used to gauge the level of influence and credibility of the leaders over a period of time.

Construction

Scoring is subjective, based on demonstrated evidence of success, and is best carried out using a 360-degree approach. You and your manager, peers, and subordinates will rate each element using a 1 to 5 range (leaving an empty score if they cannot evaluate your demonstrated performance).

There are five elements used to evaluate the CCI:

- Change History (CH)—Represents the historic experience of the leader's involvement, mentoring, and enabling prior organization's cultural and behavioral approach to achieving optimized organizational results.
- Results History (RH)—Not all changes are positive; this element evaluates the positive impacts achieved through effective organizational change management.
- Organizational Ownership (OO)—Unique to this tool, this question has a binary answer. It is either yes (scored 5) or no (scored 0) whether the change leader owns the organization.
- Organizational Influence (OI)—Influence is the polar opposite of ownership; the leader can have no ownership but great influence on the organization's transformation. This is typically scored high for effective subject matter experts who have elevated to leadership positions.
- Conviction for Change (CC)—Passion and conviction for change is one of the bedrock requirements for an effective leader to cause significant transformation. The organization must see, hear, and feel the passion of the leader. They must be able to openly receive the leader's conviction that the transformation and changes are essential to long-range organizational success.

Additionally, each element has a weight assigned. In some organizations, when they are first introducing the instrument, the elements are weighted so organizational ownership and results history carry a greater value than influence, change, and conviction. In more mature organizations, these weightings shift toward organizational influence, change history, and conviction for change. These weighting values must reflect the leadership's understanding and position on what will give the greatest value in the indexed score.

Mathematically, the CCI is the sum of each element times its prescribed weight. Equation (A.1) provides a simple construction.

$$CCI = (CH * W_{CH}) + RH * W_{RH}) + (00 * W_{00}) + (OI * W_{OI}) + (CC * W_{CC}) \quad \text{(A.1)}$$

Use a weighted approach as follows:

- Weight: Rating from 0 to 100 (relative scores must total 100)
- Score: Rating from 0 to 5 (0 = no evidence present, 5 = pervasive evidence over time)
- Results: Score × weight

Scoring Range and Definitions

Table A.2 sets the context for scoring of the five components of the index. Unique to the scoring, one component has a binary score—organizational ownership. This instrument assumes either you own the organization you are trying to change or you do not. There is no partial ownership. While this holds true for the vast number of organizations, there are examples where larger, more complex functional parts of an organization are managed using "2 in a box."

With the equation defined, the weights for each category determined, and individual scores graded, you can quickly calculate a simple column of weighted results. By adding the results together, you end up with a final weighted score. In Table A.3, an example of weighting, scoring, and calculation of results is shown.

Scoring Interpretation

What does a CCI score mean? There are several things to consider:

- Your absolute score on each element
- The score from your manager
- An average score (and its variation) from your peers
- An average score (and its variation) from your subordinates
- Comparison of how you scored yourself and each of the five above

Three things to consider any time you use the instrument are absolute level, relative level, and comparative level.

The five scoring dimensions are segmented into three different scoring analytics: absolute score, relative score, and comparative score. In each scored dimension, there are specific descriptions based on score and range that provide insight into what the scores means, how to interpret the score, and translation of the scores into action plans.

Let us break down the differences between absolute, relative, and comparative scores:

- Absolute scores are based on the granularity of the instrument. You will find the descriptions cover the full range of scores using three basic levels: scores below 2.5 (midpoint), those in the middle scoring band (2.5 to 3.5), and those in the high-scoring band, which are above 3.5.

Table A.2 Credibility Cash Index Scoring Descriptions

	Score = 1	*Score = 2*	*Score = 3*	*Score = 4*	*Score = 5*
Change History (CH)	No evidence that their organization changed through Learning with Lean	Some of their change efforts are successful at starting transformation through Learning with Lean	Many of their change efforts show successful transformation through use of Learning with Lean	Most of their change efforts result in transformation through Learning with Lean	Every organization managed learns and changes through systemic Lean improvement
Results History (RH)	No evidence of positive results from changes	Limited evidence (ad hoc) that some changes have positive results	Clear evidence that many changes have positive results	Pervasive evidence (systematic) that changes have positive results	Changes always produce positive, sustainable results
Organizational Ownership (OO)	Do not own the organization they are trying to change	*Do not score, unable to have partial ownership*	*Do not score, unable to have partial ownership*	*Do not score, unable to have partial ownership*	Own the organization they are trying to change
Organizational Influence (OI)	Cannot influence organizational change	Few examples exist of their influence in leading organizational change	Examples are evident of their influence in leading organizational change	Many examples exist where their influence resulted in organizational change	Clear, consistent influence of other organizations through the use of Lean
Conviction for Change (CC)	Do not show conviction or passion for creating change	Sometimes positive toward organizational change, passion not easily seen	Generally positive and passionate toward organizational change	Almost always show positive, passionate attitude toward organizational change	Always show, speak, and share their conviction to learn and change

Table A.3 Sample CCI Scoring

	Weight	Score	Result
Change History (CH)	15	5	75
Results History (RH)	15	5	75
Organizational Ownership (OO)	30	5	150
Organizational Influence (OI)	20	5	100
Conviction for Change (CC)	20	5	100
Total	100		500

- Relative scores are scores that provide a level of calibration when compared to the distribution of others also taking the instrument. These are broken into three scoring bands: scores below average, scores in an average scoring band, and scores above average in a high-scoring band. The importance of relative scores provides some insight on how you are rated compared with your peers. It is not unusual to find absolute scores below the middle scoring band yet your relative scores will be average. This could signify that, as a whole group, all of the people being evaluated with the tool are at a lower absolute scoring level.
- Comparative scores, on the other hand, provide discrete insight into the differences between how you are viewed by your manager, peers, subordinates, and trusted mentors. The benefit of using comparative scores provides visibility into the differences between you and your working partner groups. For example, you may have higher scores with your peers, a medium score with your manager, and low scores with your subordinates. The implications are:
  - those with whom you have a close working relationship see stronger behaviors in you than others do
  - they see stronger positive behaviors than your manager does
  - those same behaviors may be negatively viewed by your subordinates

Receiving the middle of the road score from your manager may provide an indicator that leads you to increase manager–employee communication activities. The difference between the high peer scores and low subordinate scores might prompt you to examine the exact behavior or set of behaviors you are using and adjust them accordingly as you communicate with and lead your teams.

Remember, these scores are subjective and are not statistically validated. Use caution when looking at any scored instrument as a single set of ratings. It is best to use a wider range of instruments, collect their independent feedback, and then cross-compare the findings in search of significant points of intersection. Reality is more likely to be true at the points of intersection than on any single instrument's specific feedback.

Exhibits

For copies of this exhibit's software, please contact the authors at j.zurn@hotmail.com.

Employee: ____________

	Score	Scoring Criteria				
		1	2	3	4	5
Change History (CH)		No evidence of that their organization changed through Lean learning	Some	Many change efforts show successful transformation and use of Lean learning	Most	Every organization they managed learns and changes though systemic Lean improvement
Results History (RH)		No evidence of positive results from changes	Some	Clear evidence that many changes have positive results	Many	Change always produce positive, sustainable results
Organizational Ownership (OO)		No organizational ownership	**N/A - Cannot score partial ownership**			Owns the organization they are trying to change
Organizational Influence (OI)		Cannot influence organizational change	Few	Many examples of influence in leading organizational change exist	Many	Clear, consistent influence of other organizations though the use of Lean
Conviction for Change (CC)		Does not show any conviction/passion for creating change	Slightly	Generally positive and passionate towards organizational change	Almost always	Always shows, speaks and shares their conviction to learn and change
Total:	______					

Exhibit A.4 Credibility Cash Index.

Bibliography and Works Cited

Akao, Y. (1991). *Hoshin Kanri Policy Deployment for Successful TQM.* Cambridge, MA: Productivity Press.

Argyris, C. (1977-55[5]). Double Loop Learning in Organizations. *Harvard Business Review*, 115–125.

Baldrige Performance Excellence Program. (2011). *Criteria for Performance Excellence 2011–2012.* Gaithersburg, MD: National Institute of Standards and Technology.

Barney, M., & McCarty, T. (2003). *The New Six Sigma: A Leader's Guide to Achieving Rapid Business Improvement and Sustainable Results.* Upper Saddle River, NJ: Prentice Hall PTR.

Buckingham, M., & Clifton, D. (2001). *Now, Discover Your Strengths.* New York: The Free Press—Simon & Schuster.

Burgelman, R. A. (2002). *Strategy Is Destiny: How Strategy-Making Shapes A Company's Future.* New York: The Free Press—Simon & Schuster.

Camp, R. (1989). Benchmarking: The Search for Industry Best Practices that Lead to Superior Performance. Milwaukee, WI: Quality Press.

Carroll, P. (1993). *Big Blues: The Unmaking of IBM.* New York: Crown Publishers.

Charan, R. (2001). *What the CEO Wants You to Know.* New York: Crown Publishing Group.

Chrissis, M. B., Konrad, M., & Shrum, S. (2003). *CMMI: Guidelines for Process Integration and Product Improvement.* Boston, MA: Addison-Wesley, Pearson Education Inc.

Christensen, C. M. (1997). *The Innovator's Dilemma: When New Technology Causes Great Firms to Fail.* Boston, MA: Harvard Business School Press.

Cobb, C. (2003). *From Quality to Business Excellence: A Systems Approach to Management.* Milwaukee, WI: Quality Press.

Collins, J. (2001). *Good to Great.* New York: Harper Collins.

Covey, S. (1991). *Principle-Centered Leadership.* New York: Simon & Schuster.

Crosby, P. B. (1979). *Quality Is Free: The Art of Making Quality Certain.* New York: McGraw-Hill Inc.

Deming, W. E. (1982). *Out of The Crisis.* Cambridge, MA: Massachusetts Institute of Technology.

Earle, R., Imrie, D., & Archbold, R. (1989). *Your Vitality Quotient: The Clinically Proven Program That Can Reduce Your Body Age and Increase Your Zest for Life*. New York: Warner Books.

Garwood, D., & Bane, M. (1990). *Shifting Paradigms: Reshaping the Future of Industry*. Marietta, GA: R.D. Garwood.

George, M. L., & Wilson, S. A. (2003). *Conquering Complexity in Your Business*. Dallas, TX: George Group.

Gilbert, T. F. (1996). *Human Competence: Engineering Worthy Performance*. Amherst, MA: HRD Press.

Goldratt, E. M., & Cox, J. (1984). *The Goal: Excellence in Manufacturing*. Croton-on-Hudson, NY: North River Press.

Good, M. (2003). Patient Simulation for Training Basic and Advanced Clinical Skills. *Medical Education*, 37:14–21.

Grove, A. (1983). *High Output Management*. New York: Random House.

Guaspari, J. (1988). *The Customer Connection*. New York: American Management Association.

Henderson, B., & Larco, J. (1999). *Lean Transformation: How to Change Your Business into a Lean Enterprise*. Richmond, VA: The Oaklea Press.

Hiatt, J. M. (2006). *ADKAR: A Model for Change in Business, Government and Our Community*. Loveland, CO: Prosci Learning Center Publications.

Hoerl, R., & Snee, R. (2002). *Statistical Thinking: Improving Business Performance*. Pacific Grove, CA: Wadsworth Group–Thompson Learning.

Huber, G. (1991). Organizational Learning: The Contributing Processes and the Literatures. *Organization Science*, 2(1):99–115.

Huges Aircraft Co. (1978). *R&D Productivity: An Investigation of Ways to Evaluate and Improve Productivity in Technology-Based Organizations*. Culver City, CA: Hughes Aircraft Company.

Humphrey, W. S. (2005). *PSP: A Self-Improvement Process for Software Engineers*. Upper Saddle River, NJ: Addison-Wesley.

Jones, D., & Womack, J. (2011). *Seeing the Whole Value Stream*. Cambridge, MA: The Lean Enterprise Institute.

Juran, J. (1988). *Juran on Planning for Quality*. New York: The Free Press–Macmillan.

Katzenbach, J. R., Steffen, I., & Kronley, C. (2012). Cultural Change That Sticks: Start with What's Already Working. *Harvard Business Review*, 90(7/8):110–117.

Kellerman, D. F. (1981). *New Webster's Dictionary of the English Language*. New York: Delair Publishing Company.

Kendall, J. (1990). *Combining Service and Learning, A Resource Book for Community and Public Service*. Raleigh, NC: National Society for Internships and Experimental Education.

Knerr, B. W., Lampton, D. R., Thomas, M., Corner, B. D., & Grosse, J. R. (2003). Virtual Environments for Dismounted Soldier Simulation, Training, and Mission Rehearsal: Results of the FY 2002 Culminating Event. Orlando, FL: U.S. Army Research Center.

Kotter, J. P. (1996). *Leading Change*. Cambridge, MA: Harvard Business Review Press.

Kouzes, J. M., & Posner, B. Z. (1988). *The Leadership Challenge; How to Get Extraordinary Things Done in Organizations*. San Francisco, CA: Jossy-Bass Inc.

Lareau, W. (2003). *Office Kaizen*. Milwaukee, WI: Quality Press.

Latham, J., & Vinyard, J. (2006). *Organizational Diagnosis, Design, and Transformation*. Hoboken, NJ: John Wiley & Sons.

Lencioni, P. (2002). *The Five Dysfunctions of a Team*. San Francisco: Jossey-Bass/ John Wiley & Sons.

Lewin, K. (1947). Frontier in Group Dynamics. *Human Relations*, 143–153.

Lowenstein, M. W. (1995). *Customer Retention: An Integrated Process for Keeping Your Best Customers*. Milwaukee, WI: Quality Press.

Major, G. (1992). *Take Charge! How to Manage Your Customer Relationships*. Fairfax, VA: Sigma Books.

Malandro, L. (2009). *Fearless Leadership: How to Overcome Behavioral Blind Spots and Transform Your Organization*. New York: McGraw-Hill.

Mann, D. (2005). *Creating a Lean Culture*. New York: Productivity Press.

Mascitelli, R. (2004). *The Lean Design Guidebook*. Northridge, CA: Technology Perspectives.

Mascitelli, R. (2007). *The Lean Product Development Guidebook*. Northridge, CA: Technology Perspectives.

Maslow, A. (1943). A Theory of Human Motivation. *Psychological Review*, 370–396.

Moran, R., & Riesenberger, J. (1994). *The Global Challenge: Building the New Worldwide Enterprise*. Maidenhead, Berkshire, England: McGraw-Hill International.

Ohno, T. (1988 English). *Toyota Production System: Beyond Large Scale Production*. New York: Productivity Press.

Panagiotou, G. (2003). Bringing SWOT into Focus. *Business Strategy Review*, 14(2):8–10.

Peters, T. (1985). *A Passion for Excellence: The Leadership Difference*. New York: Warner Books.

Peters, T. (1987). *Thriving on Chaos: Handbook for a Management Revolution*. New York: Alfred A. Knopf–Random House.

Prochaska, J. O., Norcross, J. C., & DiClemente, C. C. (1994). *Changing for Good: The Revolutionary Program That Explains the Six Stages of Change and Teaches You How to Free Yourself from Bad Habits*. New York: W. Morrow Inc.

Reichheld, F. (2001). *Loyalty Rules!* Boston, MA: Harvard Business School Press.

Rogers, E. M. (1995). *Diffusion of Innovations*. New York: The Free Press–Simon & Schuster.

Rother, M., & Shook, J. (2003). *Learning to See*. Cambridge, MA: The Lean Enterprise Institute.

Sandri, M. (June 2008). Signaling in Muscle Atrophy and Hypertrophy. *Physiology*, 23(3):160–170.

Scherkenback, W. W. (1991). *The Deming Route to Quality and Productivity*. Washington, DC: CEE Press Books–George Washington University.

Senge, P. (1990). *The Fifth Discipline: The Art and Practice of the Learning Organization*. New York: Doubleday.

Shewhart, W. A. (1931). *The Economics of Control of Quality of Manufactured Products*. New York: D. Van Nostrand Company Inc.

Stalk, G., & Hout, T. (1990). *Competing Against Time*. New York: The Free Press–Macmillan.

Starbird, D., & Cavanagh, R. (2011). *Building Engaged Team Performance*. New York: McGraw-Hill.

Sumanth, D. J., Edosomwan, J. A., Poupart, R., & Thor, C. G. (1995). *Productivity & Quality Management Frontiers—V.* Norcross, GA: Industrial Engineering Management Press.

Ulrich, D., & Lake, D. (1990). *Organizational Capability: Competing from the Inside Out*. New York: John Wiley & Sons.

Ulrich, D., Zenger, J., & Smallwood, N. (1999). *Results-Based Leadership*. Boston, MA: Harvard Business School Press.

Wickman, R., & Doyle, R. (1993). *Breakthrough Quality Improvement for Leaders Who Want Results*. Milwaukee, WI: Quality Press.

Index

C

D

T

U

V

W

Y

Z

About the Authors

James Zurn is an Operational Excellence consultant based in Phoenix, Arizona. He has more than 36 years of experience in quality, reliability, and design engineering with QLogic, Intel, Xerox, AT&T, and Fujitsu. Zurn holds Bachelor and Masters degrees in engineering with concentrations in statistics and operations research. He is an ASQ Certified Quality Engineer and Certified Reliability Engineer. He is an SME Certified Manufacturing Engineer in two disciplines (Manufacturing Management and Manufacturing Systems). Additionally, he is an accomplished, GE-trained, Six Sigma Black Belt. He holds registration as an ISO9000 auditor and is pursuing certification as a SEI CMMI lead assessor. He is serving his 14th year on the Board of Examiners of the Malcolm Baldrige National Quality Award as an Alumni Examiner, and is the founding chair and Judge of the Arizona Governor's Award for Quality program. He is widely published with over credits in publications such as IEEE Transactions on Engineering Management, Quality and Reliability Engineering International and ASQ's Quality Engineering.

Perry Mulligan is the senior vice president of Operations responsible for all aspects of the manufacturing and delivery of products for Emulex Corporation in Costa Mesa California. He is responsible for planning, purchasing, commodity management, quality, test and process engineering, warranty repair support as well as all IT applications and IT development for the corporation. In addition to those functional roles, Mulligan was responsible for the overall supply chain design, manufacturing strategy including the development and deployment of ODM solutions and the strategic direction for IT and technical support for customers.

He has more than 30 years of experience in operations, Supply Chain management, Commodity management, IT and manufacturing with QLogic, Solectron, Celestica, and Nortel.

Mulligan holds an MBA from the University of Western Ontario.

He is a member of the board of directors for Microvision Inc. In addition to the sitting on the board, he is also a member of the audit committee and the chair of the nominating committee.

For Product Safety Concerns and Information please contact our EU
representative GPSR@taylorandfrancis.com
Taylor & Francis Verlag GmbH, Kaufingerstraße 24, 80331 München, Germany

www.ingramcontent.com/pod-product-compliance
Lightning Source LLC
LaVergne TN
LVHW080646011026
849160LV00014B/1341

* 9 7 8 1 4 6 6 5 7 2 9 4 2 *